Principles of Medical Imaging for Engineers

Michael Chappell

Principles of Medical Imaging for Engineers

From Signals to Images

 Springer

Michael Chappell
Department of Engineering Science
University of Oxford
Oxford, UK

ISBN 978-3-030-30513-0   ISBN 978-3-030-30511-6   (eBook)
https://doi.org/10.1007/978-3-030-30511-6

This Springer imprint is published by the registered company Springer Nature Switzerland AG
The registered company address is: Gewerbestrasse 11, 6330 Cham, Switzerland

To Dad

Preface

Medical imaging is a remarkably familiar technology, being available in some form in almost all hospitals in the developed world, and increasingly worldwide. The fact that it provides a window into the living body, allowing us to 'see' things that would otherwise be impossible to look at, makes it fascinating. And, has meant it has had a significant impact on medical practice for a range of diseases and conditions. Despite our increasing familiarity and comfort with medical imaging, it is all too easy not to appreciate the differences between the different ways in which medical images are produced and thus why one technique is used in one application, but not in another.

Medical imaging technology continues to advance and the uses of it grow, but the principles on which the technology is based remain the same. As the title of this book suggests, the aim is to provide an introduction to these principles. Like a lot of subjects, there is quite a lot we could cover when describing how medical imaging works and what it is used for. We are going to restrict ourselves to the physical principles on which various imaging methods work—the 'signals'—and how we turn the information we can gather into images. The main thing that we are going to pass over, but other medical imaging books often cover, is the physics of the hardware used to acquire the signals, for example how X-ray tubes are built or ultrasound transducers work. We will overlook this because our focus will be on the principles of how we gather imaging information and thus understand what the images tell us. Once we have explored the signals and the process of generating images, we will finally look at what more we can gain from using medical imaging methods. We will see that we can do more than just generate 'pictures' but can make measurements of processes occurring in the body, something that makes medical imaging remarkably useful for studying diseases in research, as well as diagnosing diseases. If you find you need, or would like, more detailed insight into anything introduced in this book, further reading is provided for each chapter.

Fundamental to the approach taken in this book is to separate the process of generating the signals from inside the body, from the subsequent steps needed to produce the images. We will see that there are a remarkable number of ways in which we can interrogate the body to generate the signals we need. Although there are marked differences in the physics of these signal generating processes, some similar concepts will crop up repeatedly, such as attenuation, and sometimes these will help us to differentiate different tissues in the body and in other cases they cause us problems. Once we have a suitable signal, invariably more work is required to produce an image, since the signal will contain some mixture of information from different areas of the body. To actually produce images, we will meet methods for 'reconstruction' and again find similarities between the different imaging methods, but also crucial differences dictated by the way our signals were generated and the information they contain. Finally, we will, as book arising out of an engineering approach to imaging, explore how the signals and images we measure depend upon not just physical principles, but also physiological processes. And so, we will discover ways in which we can exploit medical imaging technology to make spatially resolved measurements of physiology in the body.

This book originated in a short course in medical imaging for aspiring biomedical engineers on an M.Eng. degree programme. I am indebted to the students of that course, along with the teaching assistants who tutored it, for feedback that has ultimately not only helped shape this book but also my understanding. Likewise, I am very grateful to various colleagues and collaborators who have directly or indirectly offered insight into the principles of medical imaging. In particular Tom Okell whose clear explanation of various MRI principles I hope shine through in this book in some form or another (albeit perhaps not as good as in their original form). I am also grateful to Logan, Matt, Moss and Yuriko for reading some (if not all) of the chapters and spotting various egregious errors. That said, any errors that remain in this work can only be blamed on the author.

Oxford, UK   Michael Chappell
June 2019

About This Book

By the end of this book you should:

- Understand how electromagnetic and sound waves can be used to identify structure within the body based on attenuation and reflection.
- Describe the process of tomographic reconstruction, including simple reconstruction methods such as filtered back-projection.
- Understand the concept of a point spread function and sources of error that limit the resolution of tomographic techniques.
- Describe how magnetic resonance can be used to selectively image body tissues.
- Understand why magnetic resonance imaging gives rise to signals in the frequency domain and be able to describe the resulting relationships between frequency sampling and image resolution.
- Be able to identify the contrast mechanisms exploited by the most popular medical imaging modalities.
- Be able to identify how various medical imaging signals and images depend upon physiological processes and how this can be exploited to make physiological measurements.
- Appreciate how molecules can be labelled such that they can be detected by imaging apparatus so that physiological processes can be monitored.
- Be able to quantify uptake and binding of contrast agents through the use of kinetic models.

The book includes exercises, but instead of including them all at the end of each chapter, they are deliberately inserted in the text at the most appropriate point. Many of the examples require you to explore a result that has been stated or to work through a similar example. It is not necessary to do the examples as you read through the book (and they are clearly highlighted), but the hope is that they will be helpful in reinforcing the ideas presented and provide greater insight by applying existing mathematical and physics knowledge to the topic of medical imaging. This book builds upon, and therefore presumes, knowledge of mathematics and physics that you would find in most physical science undergraduate courses. It also assumes a general knowledge of physiology and anatomy.

Contents

Part I
From Signals …

Chapter 1
Introduction

Abstract In this chapter, we consider what medical imaging is, and what makes it distinct from other forms of imaging. We then consider some principles of signals that we will need to understand the signal generation process underlying medical imaging systems.

1.1 What Is Medical Imaging?

We might regard the simplest form of medical 'imaging' as simply looking at someone—a combination of light as the imaging medium and the eye as the instrument. This is a remarkably powerful clinical method. It is possible to observe all sorts of changes in an individual just by looking at them; it is well known, for example, that people who are ill often look pale. Thus, we are able to observe physiological effects of processes at work inside the body this way. The use of photographs and videos of individuals is an active area of research for detection of diseases and monitoring of vital signs. However, at least for the purposes of this book, we will argue that this method is not 'medical imaging'. We are going to make this distinction because this method is limited to observing *surfaces*. With the eye you can only observe the outside of an individual. If you start to peer into orifices or even go as far as endoscopy you can see 'inside' the body, but again only observe the surfaces. Even with more invasive methods involving surgery, you still largely only observe surfaces. If you wanted to see inside the liver, you would have to cut into the liver—thus to see a slice through it would require physically taking a slice through it and to see all possible 'slices' would require destroying it entirely. This has been done (see the Visible Human Project[1]), but clearly this is not a practical solution for patient care.

What we want to achieve with medical imaging is to look inside someone, whilst leaving them intact, giving us complete spatially resolved information about the body. Essentially what we are after is a full three-dimensional representation of the individual or organ we are interested in. We want to be able to see inside at every point within the body and deep with the organs. What this means is that the imaging methods we are interested in will generally provide 3D information, although in

---

[1] https://www.nlm.nih.gov/research/visible/visible_human.html.

© Springer Nature Switzerland AG 2019
M. Chappell, *Principles of Medical Imaging for Engineers*,
https://doi.org/10.1007/978-3-030-30511-6_1

practice we might only be able to obtain, or at least visualize, 2D slices or 2D projections of that information.

Exercise 1A
Find out what imaging device (or devices), if any, is used in the following applications. See if you can find what reasons are given to justify the device or devices used and also what geographical differences there are in the deployments of medical imaging technology.
(a) Locating an injury in a trauma unit.
(b) Screening for abnormalities in a foetus.
(c) Discrimination between haemorrhagic and ischaemic stroke in emergency medicine.
(d) Routine screening in a doctor's surgery.

In Exercise 1A, you will hopefully have found that different imaging devices are selected for different applications, and usually there are some particular features of the imaging device that makes it particularly favourable for that application. The particular technical strengths (and weaknesses) are something we will explore further in this book. At the end you could return to the same question to provide your own justifications of what you might choose. You may also have found that more than one imaging device is used, and that the choice is often different in different countries. Inevitably cost, ease of use and even how patient-friendly imaging technology is has a role to play in when and how it is used. Thus, whilst one imaging method might be technically superior, an 'inferior' method might be used where it is cheaper and easier to do so. Partly, this will be on the principle that the lower cost and easier to use method is 'good enough' for the application. Thus, for example, whilst ultrasound produces relatively noisy images, it is increasingly widely deployed as the equipment is far more accessible to patients than MRI, even though MRI might be capable of producing far more detailed images. Safety also has an important role; thus, there is often a preference for non-ionizing methods where possible. Whilst we will touch again on some of these issues, this will not be the focus of this book, but something to bear in mind.

1.2 Signals

The first topic we are going to consider is physical principles that we can exploit for imaging. Using these principles, we are going to produce signals that are related to the tissue within the body—hence the title for this section. These signals will inevitably be different to those you might have already met if you have studied biomedical signals, i.e. largely indirect measures of physiological processes at the 'system' level, such as heart rate and blood pressure. We will consider in later chapters how we turn these signals into spatially resolved information—images.

We are going to consider the signals that lead to medical images under four categories:

- Transmission
- Reflection
- Emission
- Resonance.

This will allow us to cover all the main ways of generating the signals needed for imaging, even if we cannot cover every single example that exists within each case. However, we will start by defining some core concepts that are relevant to all imaging methods.

1.3 Noise

The basic definition of noise is unwanted signal. Thus, noise can arise from a large number of sources. We most usually consider the noise that arises from the measurement process itself: 'instrument noise'. In reality, 'noise' associated with motion of the person being imaged can be a much larger source of error. In the simplest case, we assume that every measurement we make is corrupted by noise, but that this noise is truly random and uncorrelated between measurements. Most commonly, we assume that our measured signal is the true signal s plus a noise component e:

$$y = s + e \qquad (1.1)$$

where the noise component is a random value drawn from a normal distribution with zero mean and a given standard deviation σ:

$$e \sim N(0, \sigma) \qquad (1.2)$$

This would describe white noise, and the standard deviation is a measure of the noise 'magnitude'. It is, of course, very common to define the signal-to-noise ratio as the ratio between some figures of the signal and noise magnitudes, e.g.

$$\text{SNR} = s/\sigma \qquad (1.3)$$

One way to overcome the noise is to make multiple measurements and, as you will show in Exercise 1B, if you use N measurements the noise reduces (and thus the SNR increases) by a factor of $\sqrt{N}$. There are some processes that we will be interested in that follow a different noise model or exhibit coherent noise where there is some correlation in the noise in either in time or in space. A good example that we will meet is speckle in ultrasound images. In general, averaging measurements subject to coherent noise will provide a less than a $\sqrt{N}$ improvement in SNR.

Exercise 1B

According to the model adopted in Eqs. (1.1) and (1.2) for a measurement with white noise, we can write the probability of observing a value y as:

$$P(y) = \frac{1}{\sqrt{2\pi}\sigma} e^{-\frac{(y-s)^2}{2\sigma}}$$

If we take N independent measurements of y, we might take the mean to get a more accurate estimate:

$$\hat{y} = \frac{1}{N} \sum_{i=1}^{i=N} y_i$$

Using the fact that the expectation of a sum of random variables is a sum of the expectations of the individual random variables, and the variance of this sum is (for independent random variables) the sum of the individual variances, shows that the SNR for $\hat{y}$ will be bigger than for a single measurement by a factor $\sqrt{N}$.

1.4 Contrast

Being able to quantify and quote the SNR might not be enough for medical imaging applications. For example, we might be very interested distinguishing between healthy and diseased tissue, e.g. a radiologist will be looking for tissue that has died (infarcted) in the brain of a stroke patient. A good SNR implies that we are able to obtain a very 'clean' image free from the corruption of the noise. However, if the *difference* in the signal magnitude is very small between healthy and diseased tissue compared to the overall signal magnitude, then even a small amount of noise will be enough to make our task more difficult. We call the difference between signals arising from different objects the contrast, just as you might in a picture:

$$C_{AB} = |S_A - S_B| \tag{1.4}$$

We might normalize this to get:

$$C_{AB} = \frac{|S_A - S_B|}{S_A} \tag{1.5}$$

We might also want to be able to quantify the contrast-to-noise ratio. There are various definitions, but a common one is:

$$\text{CNR}_{AB} = \frac{C_{AB}}{\sigma} \tag{1.6}$$

Exercise 1C

Using the definition of CNR in Eq. (1.6) and that of SNR in Eq. (1.4) show that:

$$\text{CNR}_{AB} = |\text{SNR}_A - \text{SNR}_B|$$

What implications does this result have for the ability to achieve a good CNR in terms of the SNR of the system being used.

1.5 Electromagnetic Spectrum

A lot, but not all, of the techniques we will meet are going to rely upon signals in some way. Figure 1.1 is a reminder of the normally accepted names for different parts of the spectrum, once we have looked at all the different sorts of signals used for imaging you might be surprised at the range of electromagnetic signals that we can exploit.

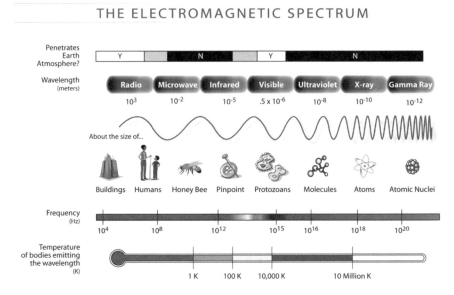

Fig. 1.1 Electromagnetic spectrum. *Source* NASA, public domain

Further Reading

For a broader introduction to medical imaging and its history see:

Medical Imaging: Signals and Systems, 2nd Ed, Jerry L Prince and Johnathan M Links, Pearson, 2015, Chapter 1.

Chapter 2
Transmission—X-rays

Abstract In this chapter, we will consider the physical principles behind transmission imaging systems. Taking X-rays as an example, we will see how attenuation can give rise to contrast between different body tissues and how scattering can affect the contrast we can achieve in practice.

For the purposes of considering 'transmission'-based methods, we are going to consider a 'light'-based, i.e. electromagnetic, technique based on the relatively high-frequency X-ray waves. This should be a very familiar method for generating images, and the chances are high that you have had an 'X-ray' or at least seen one. X-rays are most familiar from planar radiography, the conventional 'X-ray' you might get when you break a bone. However, it is also the basis of the more sophisticated 3D imaging method called computed tomography (CT). In general, in X-ray imaging we launch multiple X-rays through the body at the same time and measure them all when they emerge on the other side. We can consider the simplified scenario in Fig. 2.1 where we have an array of X-ray sources on one side and a corresponding array of detectors on the other. In practice, arrangements such as a fan beam (also shown in Fig. 2.1) are used, but the principles are sufficiently similar to the planar arrangement that we can just consider that case here.

2.1 Photon Interactions with Matter

X-rays are launched from the source through the body and measured by a detector on the other side. The main concept behind transmission methods is thus to record the different transmission of electromagnetic waves having passed through the body. X-rays are generated in an X-ray tube, the X-rays being produced by high energy electrons striking the surface of a metal target within the evacuated tube. The resulting X-rays pass through the body before arriving at the detector where they are first converted into visible light that light being subsequently detected and converted to a voltage. The detector typically measures received X-rays across a broad frequency range. To be able to use electromagnetic waves to generate medical images, we need

© Springer Nature Switzerland AG 2019
M. Chappell, *Principles of Medical Imaging for Engineers*,
https://doi.org/10.1007/978-3-030-30511-6_2

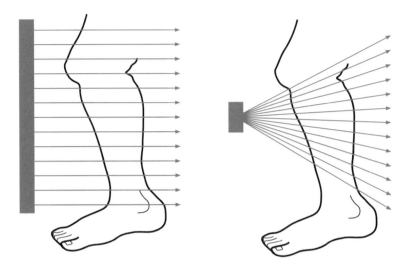

Fig. 2.1 Transmission imaging, the simplified case (left) where multiple sources give rise to parallel rays passing through the object being imaged. A more realistic scenario (right) is where the source is concentrated in a single location and a fan of beams is produced

the waves to interact with the body tissues in some way, such that the signal contains information about the tissues it has passed through. Thus, to understand what we can measure with X-rays, we need to consider the main ways in which X-ray photons can interact with matter.

2.1.1 Photoelectric Attenuation

Photoelectric attenuation is the absorption of X-rays leading to a reduction in the received signal. The first step is that an incident X-ray is absorbed by interaction with an atom, with a tightly bound electron being emitted having gained the energy carried by the photon, as shown in Fig. 2.2:

$$\text{photon} + \text{atom} \rightarrow \text{ion} + e^-$$

In a second step, an electron from a higher energy level in the atom will fill the 'hole' left behind, but since it will have gone through a smaller energy change than was involved in the release of the first electron, the energy recovered will be much lower than the original X-ray and is absorbed elsewhere in the tissue. The overall effect is complete absorption of the X-ray energy from the photon before it can reach the detector.

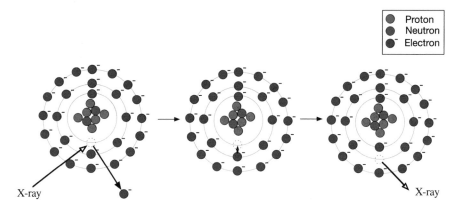

Fig. 2.2 Photoelectric interaction between an X-ray and an atom in tissue

Note that in this process, an ion is created from the atom. Hence, X-rays count as ionizing radiation. The production of ions leads to cell damage and can lead to the development of cancerous tumours where damage has occurred to DNA. For this reason, the total quantity of X-ray radiation to which an individual is exposed in a scan, the dose, is strictly limited. This also places limits on how regularly a patient might be exposed to X-rays if repeated imaging is required.

2.1.2 Compton Scattering

While the photoelectric effect dealt with the interaction of a photon with a tightly bound electron, the Compton effect refers to the interaction with more loosely bound electrons in the outer shell of an atom. In this case, only a small amount of the energy of the photon is transferred to the electron; these both cause the electron to be ejected from the atom, but also the X-ray to be deflected from its original path, as shown in Fig. 2.3. Hence, it is no longer detected by a detector placed directly opposite the source; hence, the X-ray has been attenuated as far as this detector is concerned.

Because there is only a small change in energy, the X-ray will still look like a valid signal to an X-ray detector, but as it has been deflected it will appear in the wrong place relative to those that pass straight through. Since we typically use an array of detectors, the scattered photons contribute an extra 'background' signal.

Fig. 2.3 Compton scattering of an X-ray

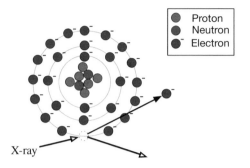

2.1.3 Other Mechanisms

There are other mechanisms by which X-rays can interact with matter; we will meet the concept of Rayleigh scattering in Chap. 3. This will not change the energy of the X-ray photon but will alter the direction of travel. Like Compton scattering, therefore, it increases the background signal in the measurements.

A particular issue, at high energies, is the secondary effects of the creation of free electrons mainly through a mechanism called pair production (where both an electron and positron are produced). However, this is only really relevant for radiotherapy, where the destructive effects on tissue are exploited to treat cancerous tumours.

2.1.4 Attenuation Coefficient

The attenuation of X-rays is typically described as an exponential process with respect to distance. For N_0 incident X-rays, the number transmitted, N, through tissue of thickness x is:

$$N = N_0 e^{-\mu(E)x} \qquad (2.1)$$

where μ is the combined attenuation coefficient from all scattering mechanisms, i.e.:

$$\mu(E) = \mu_{\text{photo}}(E) + \mu_{\text{compton}}(E) \qquad (2.2)$$

This depends upon the energy of the X-ray, E. Attention is often characterized in terms of the mass attenuation coefficient, μ/ρ (i.e. scaled by the density ρ) in units of cm^2g^{-1}. Figure 2.4 shows the mass attenuation coefficients for different tissues. As the energy of the X-ray increases, the attenuation coefficient reduces.

We can convert Eq. (2.1) into the intensity measured at the detector from X-rays passing through an object comprised of different regions of attenuation:

$$I = I_0 \varepsilon(E) e^{-\int \mu \, dl} \qquad (2.3)$$

Fig. 2.4 Typical profile for the mass attenuation coefficient for soft tissue and bone as a function of X-ray energy

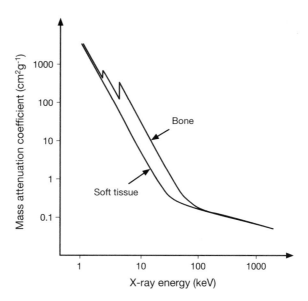

With $I_0 = EN_0$, $\varepsilon(E)$ is the sensitivity of the detector to X-rays with energy E and we have performed a line integral between source and detector to accumulate the effect of attenuation of all the different tissues along the path.

2.2 Contrast

The energy reaching the X-ray receptor will depend upon the attenuation of the X-rays along the path between source and detector. Thus, when using transmission for medical imaging, the way we encode information about tissue within the body into our signal is based on the different attenuation properties of different tissues. Hence, contrast is generated between tissues with different attention coefficients.

Looking closely at Fig. 2.4, which shows the mass attenuation coefficients for different tissues, you will see that bone has a higher attenuation than tissue, but otherwise most tissues are very similar. Hence, X-rays are good for imaging bone, but poor for creating *contrast* between soft tissues. The attenuation for bone has clearly distinguishable discontinuities, called K-edges, that relate specifically to the shell binding energies of electrons in specific atoms; for bone, these are related to calcium. The result is that X-rays with high enough energy can induce a specific photoelectric effect with calcium atoms, leading to higher absorption and thus meaning that a factor of 5–8 increase in absorption of bone over soft tissue is achieved.

Fig. 2.5 Object subject to
X-ray imaging, for use in
Exercise 2A

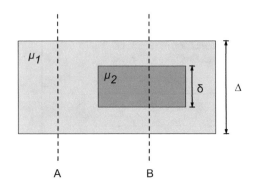

Exercise 2A

X-rays are to be used to image the object shown in Fig. 2.5 with the attenuation
coefficients given.

(a) Write down expressions for the intensity as seen by a detector located on
the lines A and B assuming that the detector sensitivity is 100%.

(b) By defining the quantity:

$$\lambda_i = -\log_e\left(\frac{I_i}{I_0}\right)$$

find an expression for the contrast in the signals measured from detectors
at A and B:

$$C = \lambda_1 - \lambda_2$$

(c) Hence, describe how this contrast measure is related to the properties of
the object.

2.3   Background Signal

Since we record the signals from multiple X-ray waves at the detector simultaneously
that have travelled through different parts in the body, scattered photons from one
wave will provide an unhelpful contribution to the signal from another wave. This
provides a background intensity to our measured signals that does not contain useful
information about the regions of different attenuation, e.g. tissues, on the X-ray path.

If we consider our detector array as a 2D surface, we can write the energy received
in an area $dxdy$ as:

$$I(x, y) = I_{\text{primary}}(x, y) + I_{\text{secondary}}(x, y) \tag{2.4}$$

We have primary photons that have arrived directly from the source but have been attenuated along their path:

$$I_{\text{primary}}(x, y) = I_0 \varepsilon(E, 0) e^{-\int \mu \, dl} \tag{2.5}$$

This follows from Eq. (2.3) but now with $\varepsilon(E, \theta)$ as the efficiency of the detector to photons with energy E received from an angle θ: for the primary photons $\theta = 0$.

We also have a contribution from secondary photons that would require us to integrate over the range of energies that the detector is sensitive to and the range of angles from which photons could be received:

$$I_{\text{secondary}}(x, y) = \int_{E_s} \int_{\Omega} \varepsilon(E_s, \theta) E_s S(x, y, E_s, \Omega) dE_s d\Omega \tag{2.6}$$

With $S(x, y, E_s, \Omega)$, the scatter distribution function, being the number of scattered photons in an energy range dE_s over a solid angle Ω. Whilst this represents a complete model of the contribution of scatter, it is too detailed to be of much practical use. Instead, it is often approximated by

$$I_{\text{secondary}}(x, y) = \left(\overline{\varepsilon_s E_s}\right) \bar{S} \tag{2.7}$$

where $\bar{S}$ is the number of scattered photons per unit area at the centre of the image (which we assume to be the worst case) and $\left(\overline{\varepsilon_s E_s}\right)$ is the average energy per scattered photon. This assumes a constant scatter contribution to every location in the final image. We can also define the scatter-to-primary ratio R, so that:

$$I(x, y) = N\varepsilon(E, 0) E (1 + R) e^{-\int \mu \, dl} \tag{2.8}$$

Exercise 2B

For the object in Exercise 2A, show that the relative contrast in intensity between measurements at A and B:

$$C = \frac{I_1 - I_2}{I_1}$$

can be given by:

$$C = \frac{1 - e^{(\mu_1 - \mu_2)\delta}}{1 + R}$$

where R is the scatter-to-primary ratio. How does the contribution of scattered photons influence the contrast between the detectors? What implications would have this for distinguishing between the two materials in the object?

2.4 Anti-scatter Grids

We have seen that scattered photons provide little spatial information, as we cannot tell where in the tissue they were scattered, but still contribute to the received intensity. Hence, they increase the background signal in the final image. This means that they reduce the contrast we can achieve between a tissue and the background itself, and also between different tissues. We can reduce their effect using an anti-scatter grid. This takes the form of parallel strips of lead foil; these strips absorb photons that do not arrive at the detector within a very tight range of angles, thus removing a large proportion of the scattered X-rays. This effectively reduces the range of integration in the calculation of secondary intensity in Eq. (2.6). However, the downside is that less energy is received overall, and thus, a larger X-ray intensity will be required, resulting in a larger dose of X-ray energy to the patient.

Exercise 2C
Which parameter in the expression for contrast that you derived in Exercise 2B is directly affected by the use of an anti-scatter grid?

2.5 The CT Number

In practice, whilst X-ray systems, such as CT, rely upon different parts of the body having different attenuation coefficients, it is not the attenuation coefficient that is usually quoted, rather it is the CT number, in which water is treated as a reference:

$$\text{CT} = \frac{1000\left(\mu - \mu_{H_2O}\right)}{\mu_{H_2O}} \tag{2.9}$$

The CT number is quoted in Hounsfield units that vary between -1000 and $+3000$; some typical values are given in Table 2.1. You will spot that bone has the highest CT number, and most tissues are similar to water.

Table 2.1 CT numbers for different tissues at 70 keV

Tissue	CT number (Hounsfield units)
Bone	1000 to 3000
Blood	40
Brain (grey matter)	35 to 45
Brain (white matter)	20 to 30
Muscle	10 to 40
Water	0
Lipid	-50 to -100
Air	-1000

Fig. 2.6 Section of tissue containing suspected tumour, for use in Exercise 2D

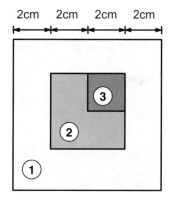

Table 2.2 Tissue properties for Exercise 2D

Tissue	Mass attenuation coefficient at 70 keV ($cm^2 \ g^{-1}$)	Density ($g \ cm^{-3}$)
1	0.25	1
2	0.3	2
3	0.3	1.5

Exercise 2D

A section of tissue containing suspected cancerous tumours is to be imaged as shown in Fig. 2.6. The component tissues possess the properties given in Table 2.2; the aim is to distinguish between tissues 2 and 3. The section of tissue is to be imaged using X-rays of energy 70 keV from a source that produces parallel X-rays placed on one side of the tissue section and detected at an array of detectors placed on the other.

(a) Sketch the quantity

$$\lambda(y) = -\log_e\left(\frac{I(y)}{I_0}\right)$$

as a function of position on the detector array y, indicating salient values.
(b) Calculate the contrast between tissues 2 and 3 as seen from your sketched profile from part (a).

Further Reading

For more information on the use of X-rays in imaging and projection-based image generation, see:

Introduction to Medical Imaging, Nadine Barrie Smith and Andrew Webb, Cambridge University Press, 2011, Chapter 2.
Medical Imaging: Signals and Systems, 2nd Ed, Jerry L Prince and Johnathan M Links, Pearson, 2015, Chapters 4 and 5.
Webb's Physics of Medical Imaging, 2nd Ed, M A Flower (Ed), CRC Press, 2012, Chapter 2.

Chapter 3
Reflection—Ultrasound

Abstract In this chapter, we will consider the physical principles behind reflection imaging systems. Taking ultrasound as an example, we will consider how differences in the acoustic properties of tissue give rise to reflections that can be used to identify interfaces within the body. We will consider how attenuation reduces the signal we can measure and scattering can be both helpful and problematic for imaging applications.

Under the topic of reflection, we are going to consider ultrasound as an example imaging signal. Sound is a mechanical pressure disturbance which propagates as a result of localized particle oscillations, giving rise to perturbations in pressure, density and particle velocity. Ultrasound relates to sounds beyond the range of human hearing and generally refers to oscillations in the range 1–10 MHz (the upper limit of human hearing is around 22 kHz).

Unlike electromagnetic radiation, which can propagate through a vacuum, sound waves require a medium to propagate through. For medical applications, this medium is body tissue and it is specific properties of this propagation that we exploit. Ultrasound may propagate both as longitudinal and/or transverse waves. In biomedical applications, ultrasound waves are normally longitudinal, as they propagate in fluid or semi-fluid media (excluding bone). Being longitudinal waves, the particles vibrate in the same direction as that of propagation of the wave.

The basic set-up for ultrasonic investigation of tissue is to use a transducer to generate an ultrasonic signal that is launched into the tissue. The transducer is made of a piezoelectric material that converts an electrical signal into a mechanical oscillation: the ultrasound wave that is launched into the body. The same device is then used to listen to the sound that is reflected back from within the tissue; the transducer converting the mechanical oscillation back into an electrical signal. This is in contrast to the transmission approach we considered in Chap. 2, where we measured the intensity of X-rays that had been transmitted through the body.

© Springer Nature Switzerland AG 2019
M. Chappell, *Principles of Medical Imaging for Engineers*,
https://doi.org/10.1007/978-3-030-30511-6_3

3.1 Wave Propagation

In Exercise 3A, you will derive the one-dimensional linear equation for the displacement of a particle, ζ, as a sound wave travels through a tissue

$$\frac{\partial^2 \zeta}{\partial z^2} = \frac{1}{c^2} \frac{\partial^2 \zeta}{\partial t^2} \qquad (3.1)$$

We can recognize this as the wave equation, and thus, we can interpret c as the speed of sound in the medium, which depends upon the (resting) tissue density ρ_0 and compressibility κ:

$$c = 1/\sqrt{\rho_0 \kappa} \qquad (3.2)$$

This is around 1540 m s^{-1} in most soft tissues, and some typical examples are given in Table 3.1. Compressibility has units of Pa^{-1} and is the inverse of the tissue bulk modulus, K. Note that Eq. (3.2) implies that the more rigid the tissue, the faster the speed of propagation and vice versa.

Exercise 3A

Derive the wave equation for tissue displacement starting with Fig. 3.1, which shows an element of tissue (at location x with length δx) with constant cross-sectional area S into the page, and its displacement when subject to a driving force, F, from the adjacent tissue. You will need to apply Newton's first law to the forces acting on the element and use Hooke's law to relate stress to volumetric strain via the bulk modulus, κ (c.f. Young's modulus). Note that in this example the cross-sectional area remains constant.

Table 3.1 Acoustic properties of biological tissues

	Acoustic impedance $\times 10^5$ (g cm^{-2} s^{-1})	Speed of sound (m s^{-1})
Air	0.0004	330
Blood	1.61	1550
Bone	7.8	3500
Fat	1.38	1450
Brain	1.58	1540
Muscle	1.7	1580
Liver	1.65	1570
Kidney	1.62	1560

Fig. 3.1 Element of tissue with constant cross section is displaced by a driving force for use in Exercise 3A

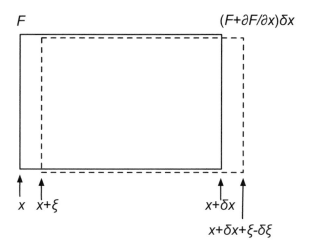

F $(F+\partial F/\partial x)\delta x$

x $x+\xi$ $x+\delta x$

$x+\delta x+\xi-\delta\xi$

The wave equation can also be used to describe the wave in terms of pressure p or particle velocity u_z:

$$\frac{\partial^2 p}{\partial z^2} = \frac{1}{c^2}\frac{\partial^2 p}{\partial t^2} \tag{3.3}$$

$$\frac{\partial^2 u_z}{\partial z^2} = \frac{1}{c^2}\frac{\partial^2 u_z}{\partial t^2} \tag{3.4}$$

You may already be familiar with solutions to the wave equation and in particular D'Alembert's solution which includes both a forward and backward travelling wave:

$$p(z,t) = F(z - ct) + G(z + ct) \tag{3.5}$$

Ultrasonic systems can exploit both continuous waves (CW) and pulses. Pulsed ultrasound can be modelled in a similar way to CW, but the finite length of the pulses must be taken into account. By nature of being limited in time, pulses contain a range of frequencies, so the key quantity is the centre frequency.

Like any wave, there are two ways of measuring the speed at which pulsed ultrasound propagates:

- Phase velocity: each frequency component in the pulse will travel at its own phase velocity c.
- Group velocity: the velocity at which the centre of the pulse propagates, evaluated at the centre frequency.

In a dispersive medium, the variation in phase velocity with frequency will result in the pulse becoming distorted with distance.

Intensity is the most commonly used measure of energy in a wave because it relates directly to human perception of sound, i.e. loudness. Intensity, I, is normally quoted in W m^{-2} and equals *the rate of energy transferred per unit area in the direction of propagation*. For a plane sinusoidal wave with pressure amplitude p, propagating in a medium of density ρ and phase velocity c:

$$I = \frac{p^2}{\rho c} \tag{3.6}$$

The range of intensities encountered in acoustics covers many orders of magnitude. Thus, energy is often expressed in terms of an intensity level, IL, in decibels (dB): the log ratio of a given intensity with respect to reference intensity (normally the threshold of human hearing $\sim 10^{-12}$ W m^{-2} at 1 kHz):

$$\mathrm{IL(dB)} = 10 \log_{10}\left(\frac{I}{I_{\mathrm{ref}}}\right) \tag{3.7}$$

For ultrasound-based medical signals where we listen to the reflected echoes from inside the body, we are most often interested in measured intensity relative to that generated by the source.

3.2   Acoustic Impedance

An important parameter for ultrasonic waves is the specific acoustic impedance, Z. You may have met the idea of impedance before in electronics as the ratio of voltage and current (c.f. Ohm's law). In general, it can be thought of as the ratio of a 'force' to a 'flow'. For sound waves, Z is defined as the ratio of acoustic pressure to particle velocity:

$$Z = \frac{p}{u_z} \tag{3.8}$$

Note that like other examples of impedance you may have met, the specific acoustic impedance can depend upon frequency and take a complex value. However, for the plane waves that we will consider here, Z is not frequency dependent and also is purely real, and thus, *the pressure is in phase with the particle velocity*. This allows us to define the characteristic acoustic impedance, which is a property of the tissue only:

$$Z = \rho_0 c = \sqrt{\frac{\rho_0}{\kappa}} \tag{3.9}$$

With units of kg m^{-2} s^{-1}. Typical values for this are given in Table 3.1. We will henceforth simply use the term impedance for acoustic impedance, but the term 'acoustic impedance' is also used for the ratio of pressure to volume velocity.

3.3   Reflection and Refraction

Figure 3.2 shows the behaviour of an incident sound wave when it hits a boundary with different acoustic properties: both a reflected and transmitted component results. Note that this assumes that the boundary is larger than the ultrasound wavelength. The standard results apply, mirroring what is seen with light waves at a boundary, namely

$$\theta_i = \theta_r \tag{3.10}$$

$$\frac{\sin \theta_i}{\sin \theta_t} = \frac{c_1}{c_2} \tag{3.11}$$

As you will have seen from Table 3.1, the wave speed is similar for most body tissues, and thus, the angle of incidence and transmission is typically approximately the same. The exceptions are interfaces between tissue and air or bone, both of which would cause refraction of the ultrasound wave.

It is possible to derive reflection and transmission coefficients, describing the proportion of the wave that is transmitted and that which is reflected, for the general case in Fig. 3.2. However, we are mostly going to be interested in the case where the angle between the incident wave and the boundary is 90°. The following results give the reflection and transmission coefficients for both pressure and intensity for this simple case:

$$R_p = \frac{Z_2 - Z_1}{Z_2 + Z_1} \tag{3.12}$$

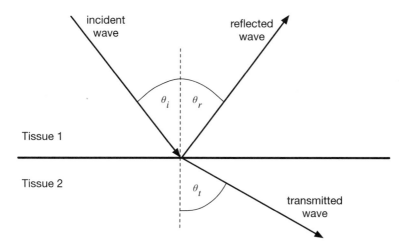

Fig. 3.2   Behaviour of an ultrasound beam at a boundary between tissues with different acoustic properties

$$T_p = \frac{2Z_2}{Z_2 + Z_1} \tag{3.13}$$

$$R_I = \frac{(Z_2 - Z_1)^2}{(Z_2 + Z_1)^2} \tag{3.14}$$

$$T_I = \frac{4Z_1 Z_2}{(Z_2 + Z_1)^2} \tag{3.15}$$

Note that in ultrasound imaging it is R that we mainly care about since we want to use the reflections to detect boundaries in the body arising from different tissues. Although, whatever is transmitted through one interface provides the signal needed to generate a reflection from interfaces deeper within the body.

The reflected signal will be maximal if either Z_1 or $Z_2 \to 0$. Under this situation, the ultrasound wave would never reach deeper structures, so is not ideal for imaging applications. This is typical if there is a tissue/air interface in the region being imaged: the air space creates a 'shadow' region behind it which cannot be imaged. The large impedance difference between tissue and air is also the reason why a coupling gel is required to eliminate air between imaging probe and skin. We could consider this a case of impedance matching as might be familiar from electromagnetic waves and communication electronics.

If $Z_1 = Z_2$ then $R = 0$, there is no reflection and we cannot see the boundary at all. From Table 3.1, you can see that Z are similar for most soft tissues, and thus, the reflected component from the boundary will be small, of the order of 0.1% intensity.

3.4   Scattering

If the ultrasound wave strikes an object that is a similar size or smaller than the wavelength, then it gets scattered in all directions. The variation in angular intensity of the scattered wave will depend upon shape, size, physical and acoustic properties of the scattering object. If the size of the scatterer is small compared to the wavelength, then scattering is relatively uniform in direction—with slightly more energy being scattered backward. This is named Rayleigh scattering, and the scattered energy rises with the fourth power of frequency. Where the scatterers are densely packed, the scattered waves can add constructively. This occurs when ultrasound meets red blood cells in a vessel and gives rise to the ability of ultrasound to measure blood flow. If the scatters are more widely and randomly spaced, then the effect is to introduce 'noise' into the images called speckle.

3.5 Absorption

We have already seen that as ultrasound passes into the body its energy will be attenuated by reflection and scatter. The intensity is also reduced by absorption, and this can occur by a number of different processes. The most important is relaxation absorption. When the ultrasound wave passes through the tissue, it displaces the tissue. Body tissues are largely elastic media, and thus, the displaced tissue experiences an elastic force to return to its equilibrium position, and the time taken for this to occur will depend upon the physical properties and can be described by relaxation time. If the relaxation coincides with the force exerted by the ultrasound wave, then relatively little energy will be extracted from the wave—as both elastic forces and the force exerted by the ultrasound are pulling in the same direction. However, if the restoring forces in the tissue are trying to return the tissue to equilibrium at the same time as the ultrasound wave is trying to force the tissue away from equilibrium, then a larger amount of energy will be lost. This is essentially a resonance process: when the tissue's characteristic frequency (inverse of the relaxation time constant) matches that of the ultrasound wave then very little absorption will occur, but otherwise energy will be lost.

The relaxation process can be characterized by an absorption coefficient:

$$\beta_r = \frac{\beta_0 f^2}{1 + \left(\frac{f}{f_r}\right)^2} \tag{3.16}$$

This is a function of the ultrasound frequency, f, and highly dependent upon the relaxation frequency of the tissue, f_r. In reality, the body contains a range of tissue with different f_r and the overall absorption coefficient is the sum of these. Measured values of absorption coefficient in many tissues have found an almost linear relationship between it and ultrasound frequency.

Another, less important, mechanism for absorption in the body is called classical absorption: caused by friction between particles. It can also be characterized by an absorption coefficient, in this case one that is proportional to the square of ultrasound frequency.

3.6 Attenuation

Attenuation of the ultrasound wave is thus a combination of absorption and scattering. The general law for intensity at a point x within a (uniform) tissue is:

$$I(x) = I_0 e^{-\mu x} \tag{3.17}$$

where μ is the intensity attenuation coefficient of the tissue. This is the same model as we used for X-rays in Chap. 2. The units of μ are m^{-1}, but it is usual to quote them in nepers cm^{-1} (nepers is a non-dimensional unit reflecting the fact that we are using base e—the Naperian logarithm). If the wave is passing through tissue with a range of different values of attenuation coefficient, we integrate along the path of the wave (in exactly the same way as we did for X-rays):

$$I = I_0 e^{-\int \mu dl} \tag{3.18}$$

Practically, intensity ratios are expressed in decibels (thus using base-10):

$$\mu[\text{dB cm}^{-1}] = 4.343\mu[\text{cm}^{-1}] \tag{3.19}$$

Exercise 3B
(a) Starting with Eq. (3.17) shows that the relationship between intensity attenuation coefficient in dB cm^{-1} and cm^{-1} is as given in Eq. (3.19).
(b) Attenuation can also be expressed in terms of pressure amplitude:

$$p = p_0 e^{-\alpha x}$$

where α is the amplitude attenuation coefficient. Using Eq. (3.6) shows that $\mu = 2\alpha$.
(c) Hence, show that when expressed as a ratio in dB cm^{-1} the two attenuation coefficients take the same value:

$$\alpha[\text{dB cm}^{-1}] = \mu[\text{dB cm}^{-1}]$$

Attenuation coefficients will vary with frequency, and a simple model is often adopted:

$$\mu\left(\text{dB cm}^{-1}\right) = Af^n \tag{3.20}$$

Some typical values for A and n are given in Table 3.2. For most biological tissues, the relationship with frequency is linear (i.e. $n \approx 1$) and $A \approx 1$.

3.7 Ultrasonic Contrast

We have now seen that there are a variety of properties of the tissues that will affect the magnitude of the measured ultrasound signal: speed of sound, impedance and attenuation. For transmission methods, like X-rays, it was the attenuation properties

Table 3.2 Some typical attenuation properties of various body tissues

Tissue type	A (dB cm^{-1} MHz^{-1})	n
Water	0.00217	2
Blood	0.14	1.21
Brain	0.58	1.3
Breast	0.75	1.5
Bone (skull)	6.9	1
Fat	0.6	1
Heart	0.52	1
Kidney	0.32	1
Liver	0.45	1.05
Muscle	0.57–1.3	1
Spleen	0.4	1.3

of the tissue that we used to generate contrast. For reflection techniques, it is the impedance mismatches that are more useful, allowing us to measure reflected signals from the different interfaces and interpreting the magnitude of the reflected signal as a contrast between them based on the differences in impedance between the two tissues. We will see in Chap. 7 how we can separate the different reflections and thus produce images from the signals.

For this form of image signal generation, attenuation is a nuisance, and it reduces the measured signal magnitude, but without providing any additional information about the tissue. For ultrasound, it is attenuation that means that in practice we cannot use a transmission approach, and most parts of the body are too thick for ultrasound to pass through. Attenuation also limits the depth penetration of ultrasonic imaging. The frequency dependence of the attenuation coefficient means that lower frequencies, e.g. 2 MHz, have to be used where we want to image deeper structures, such as in the abdomen.

Exercise 3C

A section of tissue is to be imaged using ultrasound at 2 MHz. The arrangement of different tissues is shown in Fig. 3.3, and these possess the properties in Table 3.3. Assuming an attenuation coefficient of 0.5 dB cm^{-1} MHz^{-1} and a speed of sound of 1540 ms^{-1} for all tissues (and the coupling gel), calculate the intensity of the ultrasound received at the transducer relative to that transmitted for each of the tissue interfaces.

(a) Calculate the intensity of ultrasound that reaches the far side of the tissue.

(b) Repeat this calculation for ultrasound of 8 MHz and comment on the relative depth penetration of different ultrasonic frequencies.

Fig. 3.3 Section of tissue,
for use in Exercise 3C

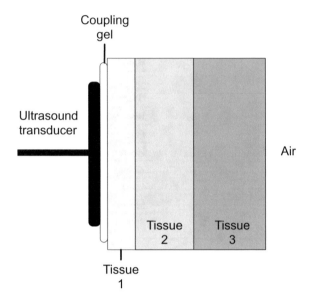

Table 3.3 Tissue properties

Tissue no.	Thickness (cm)	Acoustic impedance $(g\ cm^{-2}\ s^{-1})$
Gel	0.1	1.0
1	0.5	1.4
2	1.0	1.6
3	1.5	8.0

3.8 Doppler Ultrasound

We can exploit the Doppler effect experienced by ultrasound waves reflected off a moving object to calculate velocities. The most common application of this is to measure blood flow velocity using the signal from red blood cells (RBCs). Because RBCs are of the order of 10 μm, but are also fairly densely packed in the blood, the signal received from them is a scattered signal. Since scattering from RBCs directs energy in all directions the backscattered signal received by the transducer is small and low intensity. The signal intensity is proportional to the fourth power of the ultrasound frequency, so higher frequencies are typically used for blood velocity measurements.

Exercise 3D

(a) Using the geometry in Fig. 3.4 and noting that the component of the velocity of RBCs towards the transducer is $v\cos\theta$, show that the 'apparent' frequency of the ultrasound wave as seen by the RBCs is:

$$f_i^{\text{eff}} = \frac{c + v\cos\theta}{\lambda} = f_i\frac{c + v\cos\theta}{c}$$

(b) Hence, find an expression for the apparent frequency of the received wave.

(c) By defining the difference in the transmitted and received frequencies:

$$f_D = f_i - f_r$$

and using the fact that $v \ll c$, find an expression for Doppler frequency.

As you will have shown in Exercise 3D, the Doppler shift is:

$$f_D \approx \frac{2f_i v\cos\theta}{c} \tag{3.21}$$

where f_i is the frequency of the ultrasound being used, $v\cos\theta$ is the velocity of the blood expressed as a speed and angle relative to the ultrasound wave direction. If

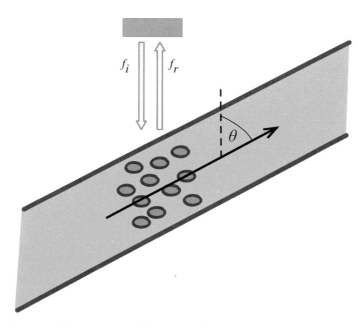

Fig. 3.4 Doppler shift measurement of blood velocity

we measure this effect at the transducer, we can estimate the velocity. Note that this is a true measure of velocity, or at least the component in the direction parallel to the wave propagation direction, and not just speed, since we can tell the difference between RBCs travelling towards and away from the transducer based on whether the Doppler shift is positive or negative. Note that aliasing could cause errors in Doppler measurements at high flow velocities.

Exercise 3E

A 5 MHz ultrasound device is to be used to measure blood flow speed using the Doppler effect.

(a) If a Doppler shift of 2.26 kHz is measured, calculate the flow speed for a blood vessel oriented at 45° to the transducer. What fractional change in ultrasound frequency is this?

(b) What further information about the Doppler shift would you need to calculate flow *velocity*.

Further Reading

For more detail on ultrasound, propagation and reflection see:

Introduction to Medical Imaging, Nadine Barrie Smith and Andrew Webb, Cambridge University Press, 2011, Chapter 4.
Medical Imaging: Signals and Systems, 2nd Ed, Jerry L Prince and Johnathan M Links, Pearson, 2015, Chapter 11.
Webb's Physics of Medical Imaging, 2nd Ed, M A Flower (Ed), CRC Press, 2012, Chapter 6.

Chapter 4
Emission—SPECT/PET

Abstract In this chapter, we will consider the physical principles behind emission imaging systems. Taking single-photon emission and positron emission as examples, we will consider how radioactive emissions can be used to generate signals associated with the presence of tracer materials introduced into the body. We will consider how attenuation and scattering affect the signal we measure.

Emissive techniques generally rely on using radioactive materials to generate emissions inside the body that can be detected externally. One of the key advantages of this approach is that body tissue naturally has very low radioactive emissions, which means that the background signal, e.g. from tissue we are not interested in, is practically zero. However, since there is no inherent signal from the body, we have to do something to get the emissive material into the right place. Thus, these techniques rely on the injection of a radiotracer (radioactive tracer), an agent that is usually specifically designed or chosen to be absorbed by cells or bind to a specific site in the body. We will consider radiotracers further, along with their kinetics, how these tracers behave, in Chaps. 11 and 12.

4.1 Radionuclides

There are two major mechanisms exploited in medical imaging using emission techniques, and these give rise to the names used for the associated imaging methods: single-photon emission computed tomography (SPECT) and positron emission tomography (PET).

4.1.1   Single-Photon Emission

We can exploit unstable nuclei that emit gamma radiation as shown in Fig. 4.1. This is a radioactive process in which the nucleus remains unaltered in terms of the number of subatomic particles:

$$_{Z}^{A}X_N \rightarrow {}_{Z}^{A}X_N$$

The signal we measure is gamma wavelength electromagnetic energy. Some typical single-photon emitters are shown in Table 4.1, notice that the energy in the photons they emit varies dependent upon the nucleus in question.

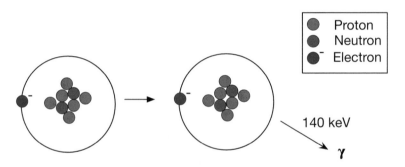

Fig. 4.1   Single-photon emission

Table 4.1   Common SPECT and PET radionuclides

Radionuclide	Maximum positron energy (MeV)	$T_{1/2}$	Photon energy (keV)
Single-photon emitters			
^{99m}Technetium	–	6.0 h	140
^{123}Iodine	–	13.0 h	159
^{133}Xenon	–	5.3 days	80
Positron emitters			
^{15}Oxygen	1.72	2.1 min	511
^{13}Nitrogen	1.19	10.0 min	511
^{11}Carbon	0.96	20.3 min	511
^{18}Flourine	0.64	109.0 min	511

4.1.2 Positron Emission

The alternative is to use positron emitters as shown in Fig. 4.2. Positron emission is a subtype of beta decay in which a proton inside the nucleus is converted into a neutron whilst releasing a positron and an electron neutrino:

$$_Z^A X_N \rightarrow \, _{Z-1}^{A} X_{N+1} + e^+ + \nu$$

Positron emission methods do not rely upon the detection of the positrons themselves, and these actually only travel a short distance before annihilating with an electron, as in Fig. 4.2. This process produces two gamma photons:

$$e^+ + e^- \rightarrow \gamma + \gamma$$

These photons both have 511 keV energy, higher than in single-photon emission, and are emitted at precisely 180° from each other. Typical positron emitters are shown in Table 4.1.

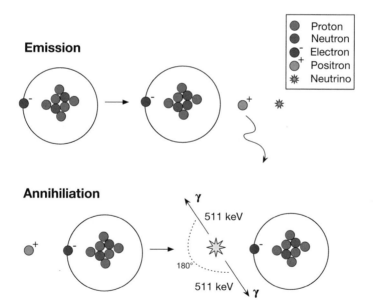

Fig. 4.2 Positron emission and annihilation

4.1.3   Radioactive Decay

With all these radionuclides, the half-life of decay is important, and some of these are given in Table 4.1, as this determines how long we have to image the subject before the radioactivity is no longer measurable. SPECT tracers tend to be generated on-site from a radioactive source with a longer half-life, making them reasonably practical for medical applications. PET radionuclides need to be generated in a synchrotron, and these are typically large and expensive devices and not widely available. Imaging with radionuclides with a short half-life, such as oxygen, can generally only be done at a site with a synchrotron. Longer-lived radionuclides can be generated in one place and shipped for use at another site, and thus, ^{18}F is the most common PET positron emission nuclide.

4.2   Detection

Both SPECT and PET rely upon the detection of gamma photons, albeit with different energies, as indicated by the values in Table 4.1.

4.2.1   Gamma Camera

Detection is done using a gamma camera which consists of a scintillation crystal and coupled photomultiplier tubes. When a γ-ray strikes the crystal, it loses energy through photoelectric and Compton interactions with the crystal (which we met in Chap. 2 when looking at X-rays). The electrons ejected by these interactions in turn lose energy through ionization and excitation of the scintillation molecules. As these molecules return to their unexcited state, they emit photons, for the most common detector, using thallium-activated sodium iodide, emission occurs at a wavelength of 415 nm (visible blue light). The photomultiplier tube, in which a photon striking a cathode gives rise to the release of a large number of electrons, effectively amplifies the signal. Although the photomultiplier tubes themselves are quite large, several centimetres across, better localization of individual events is achieved by capturing both the large current in the nearest photomultiplier tube and smaller signals that occur in nearby tubes and making assumptions about the relationship between signal and distance from the event.

   The gamma camera used for SPECT will include a collimator that has a similar design to anti-scatter grid in X-ray imagining. Unlike an X-ray system, where photons are launched into the body from a specific direction, in SPECT they are emitted in all directions from source atoms within the body, and thus, a much higher degree of collimation is required if we want to only acquire photons from a specific direction, resulting in over 99.9% of the gamma rays not reaching the detector. The spacing of

Fig. 4.3 Collimator arrangement for use in Exercise 4A

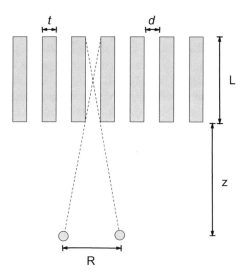

the septa in the collimator will thus ultimately determine the spatial resolution of the gamma camera. For PET, the collimator is not required as we can make use of the paired gamma photons to determine the direction from which the gamma photons arise, something we will return to in Chap. 8.

Exercise 4A
(a) For the collimator arrangement in Fig. 4.3, determine the relationship between the maximum separation of two-point sources, R, and the parameters of the collimator that would lead to them being interpreted as a single object.
(b) Comment on the effect of collimator design on the ability to resolve separate objects in the final image.
(c) Why will the SPECT collimator not improve the SNR (and thus CNR) in the same way that an anti-scatter grid would in X-ray detection.

4.2.2 Attenuation

Like the other methods based on electromagnetic radiation, we need to consider attenuation of the γ-rays produced as part of both the SPECT and PET methods. As in Chap. 2, we can assume attenuation follows an exponential model with attenuation coefficient. Like with X-rays, attenuation will arise through a combination of absorption and scattering.

For emission methods, we are interested in measuring the radionuclide concentration since it is the presence of the radionuclide, and often the concentration of the tracer itself, in a given body region, that supplies the contrast we need in the resulting images. Thus, attenuation of the signal is a nuisance, and we need to correct for it. The simplest correction is to assume uniform attenuation coefficient and then correct based on an estimate of depth of tissue which the γ-rays have passed through. Since individual signals do not contain any information about the distance between emission and detector, we have to use information from the tomographic reconstruction for this correction, something we will consider in Chap. 8. This correction ignores important tissue-dependent variations in attenuation. Commonly these days SPECT and PET devices are combined with a CT scanner, allowing a CT image to be used to capture variations in tissue across the body, and this is converted to a map of attenuation coefficient based on typical tissue properties.

Exercise 4B
For the SPECT radionuclide ^{99m}Technetium (with energy 140 keV), the attenuation coefficient is 0.145 cm^{-1}, what difference in signal intensity due to attenuation would be observed when imaging in the abdomen.

4.2.3   Scatter

Scatter is an issue for SPECT and PET systems as it is for transmission systems. The detectors used are typically sensitive over a relatively wide range of energies, which means that a gamma ray that has been deflected even over a relatively large angle will still be detected even though its energy will have changed. Whilst the collimator in SPECT will reduce the scatter energy received, it does so in exactly the same way as for non-scattered rays. Thus, the contribution from scattered gamma rays in SPECT is similar to those that have not been scattered and will provide background intensity to all of the signals that will tend to reduce the ability to distinguish true signal and thus diminish contrast in the final image. For PET, the way that scattered photons influences the final signal is a bit different, something we consider in Chap. 8, but the ultimate effect is similar.

The most common correction method uses a dual (or multiple) energy window approach, whereby the energy received by the detector is divided into multiple different ranges of gamma ray energy: 'windows'. The primary window around the energy of the gamma rays from the source is reconstructed and separately the contribution from another, secondary, window that should only contain scattered energy is also computed. With suitable scaling, the scatter contribution can be subtracted from the primary. Alternatively, for PET/CT the CT image might be used to compute maps of the expected scatter contribution which can then be included in the image formation process.

4.2.4 Deadtime

Deadtime refers to the finite recovery time in the detector that prevents another photon from being recorded. This usually means that as the true count rate increases the measured count rate saturates towards a maximum value. Correction for this is usually based on knowledge of the characteristics of the system.

Further Reading

For more information on single-photon and positron emission systems see

Introduction to Medical Imaging, Nadine Barrie Smith and Andrew Webb, Cambridge University Press, 2011, Chapter 3.
Medical Imaging: Signals and Systems, 2nd Ed, Jerry L Prince and Johnathan M Links, Pearson, 2015, Chapters 7 and 8.
Webb's Physics of Medical Imaging, 2nd Ed, M A Flower (Ed), CRC Press, 2012, Chapter 5.

Chapter 5
Resonance—Nuclear Magnetic Resonance

Abstract In this chapter, we will consider the principles of nuclear magnetic resonance to see how it can be exploited to generate signals from within the body related to the presence and properties of water within tissues. We will explore how radiofrequency waves can be used to exploit resonance phenomenon of protons placed within a magnetic field and the various timing parameters that are associated with the relaxation of the resonant system.

Resonance will probably be the least familiar way of generating signals from inside the body that we will look at in this book. Strictly, what we are going to talk about is nuclear magnetic resonance (NMR) imaging. Although the imaging community has dropped the 'nuclear' part of the title and it is now always described as magnetic resonance imaging (MRI). This helps to distinguish it from nuclear decay in the 'nuclear imaging' methods of SPECT and PET and also to avoid unfortunate confusions with nuclear weapons/accidents, etc. The clues to how we exploit NMR to generate imaging signals are there in the name:

- **Nuclear**—We get signals from the atoms in our sample and specifically from the nuclei. However, as we have noted, it has nothing to do with radioactive decay like in SPECT and PET.
- **Magnetic**—It requires the presence of (very) strong magnetic fields so that the nuclei behave in the way we need them to.
- **Resonance**—We exploit a resonance condition of the nuclei in a strong magnetic field that allows us to probe the sample using radiofrequency waves.

NMR is a very widely used chemical analytic technique that can probe the different nuclei present in a sample, as well as their interaction with other nuclei via chemical bonds. It can thus tell us information about the substances present in a sample. However, strictly NMR cannot tell us where the nuclei are. Like all the other methods, we have looked at so far, we will leave discussion of how we turn the information into images until later and just look at how we generate a signal first.

A very simple description of an NMR experiment would be that a radiofrequency wave is transmitted into a sample that is sat in a magnetic field and the signal that returns (is re-emitted) from the sample is recorded.

© Springer Nature Switzerland AG 2019
M. Chappell, *Principles of Medical Imaging for Engineers*,
https://doi.org/10.1007/978-3-030-30511-6_5

NMR goes right down to the atomic level, and thus, it should strictly be necessary to use quantum theory to describe the physics. Conveniently for us, the principles we need can also adequately be explained using classical (Newtonian) physics. If you really want to know about the quantum description most specialist textbooks will include it (see Further Reading for this chapter).

5.1 Microscopic Magnetization

All nuclei with an odd atomic weight and/or atomic number possess a quantum mechanical property termed 'spin'. This can be imagined[1] as the charged nucleus spinning around an internal axis of rotation. Atomic nuclei that 'spin' have specific energy levels related to their spin quantum number, S. The number of energy states is given by $2S + 1$. For a typical nucleus possessing spin with $S = 1/2$, the number of energy states available is 2. These two energy states are denoted by $-1/2$ and $+1/2$ and are aligned in opposite directions: one pointing 'North' (= parallel), the other 'South' (= anti-parallel).

A nucleus that has spin acts like a small dipole magnet, in the same way that a spinning charged object would, as shown in Fig. 5.1. The two energy states correspond to the dipole pointing 'upward/north' or 'downward/south'. The nucleus thus possesses a microscopic magnetic field with a magnetic moment vector:

Fig. 5.1 'Classical physics' view of the 'spin' property of a nucleus

---

[1]What we are doing here is describing a 'model' of how the system behaves in a way that allows us to write down mathematical expressions and predict the behaviour. As long as this model can predict all the behaviour we will see in practice, it will be good enough for our needs.

Table 5.1 Common gyromagnetic ratio for spin 1/2 systems

Nucleus	γ	
	10^6 radians s^{-1} T^{-1}	MHz T^{-1}
^{1}H	267.51	42.58
^{13}C	67.26	10.71
^{19}F	251.66	40.05
^{23}Na	70.76	11.26
^{31}P	108.29	17.24

$$\mu = \gamma \, \Phi \tag{5.1}$$

where Φ is the angular momentum of the nucleus and γ the gyromagnetic ratio in units of radians s^{-1} T^{-1}. Thus, the size of this magnetic moment vector depends upon the strength of the magnetic field in which the nucleus is, hence the appearance of T (Telsa) in the units for the gyromagnetic ratio. It is often more convenient to express γ in MHz T^{-1}, and values for this parameter for common spin ½ systems are given in Table 5.1.

For imaging use, the most common nucleus to exploit is hydrogen because it is by far the most abundant in the body. Other nuclei might be exploited in other very specific, and largely research, applications, e.g. ^{13}C can be used for tracking metabolic products; however, it is a very rare isotope of carbon in physiological systems where you mostly find ^{12}C. Since the hydrogen nucleus is just a bare proton, it is common to refer to them simply as protons or even as 'spins', and we will often use that terminology here. The main source of protons in biological tissues is water and thus we are largely going to be generating signals (and ultimately images) that tell us about the properties of water.

5.2   Precession

When a proton is placed into an external magnetic field B_0, it will start to 'wobble' or precess around the axis of that external magnetic field. This motion is in addition to the proton's spin about its own axis. Remember that the axis about which the proton is spinning is not directly aligned with the magnetic field. This precession is a combination of the spin about the proton axis and the angle of that axis relative to the main field and is illustrated in Fig. 5.2. The classic analogy for this is a spinning top.

The Larmor equation gives the rate at which the proton precesses around the external magnetic field:

$$\omega_0 = \gamma \, B_0 \tag{5.2}$$

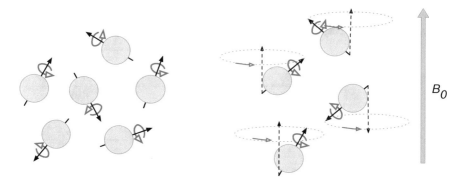

Fig. 5.2 Precession of protons in an applied magnetic field. In the absence of a magnetic field (left), orientation of the protons magnetic polarity is random. When a field B_0 is applied the protons precess around the direction of the applied field, either aligning with (lower energy) or against (higher energy) the direction of the field. Individual protons will not be in phase with each other

where ω_0 is the angular frequency of precession (in radians per second) which is called the Larmor frequency, and we have met γ already: the gyromagnetic ratio.

Exercise 5A
Many hospital MRI scanners operate at 1.5 or 3 T. Calculate the Larmor frequency associated with protons in these systems. Compare this to the frequencies used in broadcast transmission of radio and television signals.

5.3   Macroscopic Magnetization

In the absence of an externally applied magnetic field, there is no preferred direction of nuclei in a sample of material, as shown in Fig. 5.2. Thus, on a macroscopic level, the individual spins cancel each other out, and the resulting spin system has no macroscopic magnetic field. If the spin system is placed into an external magnetic field, the situation changes. The spin system becomes macroscopically magnetized, as the microscopic spins tend to align with the external magnetic field. The magnitude and direction of an external magnetic field can be represented by:

$$\mathbf{B}_0 = B_0\hat{\mathbf{z}} \tag{5.3}$$

where B_0 is the magnitude of the magnetic field in Tesla and $\hat{\mathbf{z}}$ is a unit vector pointing in the $+z$ direction, which by convention is the direction in which the magnetic field is applied.

If a spin 1/2 system is placed into the magnetic field B_0, then at equilibrium, the microscopic magnetization, μ, will be parallel to the z-axis and points in one of two possible orientations with respect to $\hat{\mathbf{z}}$:

- Parallel: μ has a component in $+\hat{\mathbf{z}}$, the precession is thus around $+\hat{\mathbf{z}}$. This is called the 'up' direction and is the low-energy state.
- Anti-parallel: μ has a component in $-\hat{\mathbf{z}}$, the precession around $-\hat{\mathbf{z}}$. This is called the 'down' direction and is the higher-energy state.

The x–y orientation of the spin around the z-axis, the phase of μ, is random. The spin system becomes slightly magnetized in the z-direction because there is a slight preference for the spins to align in the 'up' direction, as it is the more energy favourable state, as illustrated in Fig. 5.2.

We can describe the 'bulk' or macroscopic magnetization by:

$$\mathbf{M} = \sum \boldsymbol{\mu}_n = \gamma \mathbf{J} \tag{5.4}$$

where $\mathbf{J}$ is the bulk angular momentum. We have already noted that the x–y orientation of μ is random, and thus, we would see no net transverse magnetization, i.e. zero x and y components for $\mathbf{M}$. If the sample is left undisturbed for a 'long time' (we will define later what we mean by 'long' and 'short' times), $\mathbf{M}$ will reach equilibrium $\mathbf{M}_0$ which is parallel with $\mathbf{B}_0$ with magnitude M_0:

$$M_0 = \frac{B_0 \gamma^2 \hbar^2}{4kT} P_{\mathrm{D}} \tag{5.5}$$

where k is the Boltzmann constant, $\hbar$ is the reduced Planck's constant, T is the absolute temperature and P_{D} is the proton density (number of targeted nuclei per unit volume that are mobile). Obviously, M_0 becomes larger for larger B_0 or larger P_{D}, which implies that if we could measure M we might be able to say something about the proton density, and in turn, this might be a way to generate a contrast between different tissues.

Boltzmann's equation yields the relative number of protons in each of the two configurations:

$$\frac{N_{\text{anti-parallel}}}{N_{\text{parallel}}} = e^{-\frac{\Delta E}{kT}} \tag{5.6}$$

where $\Delta E = \gamma \hbar B_0$. The energy difference ΔE between the two energy states depends on B_0 and determines the relative number of protons in either state. In a 3 T field, which is relatively typical clinical field strength, there are only about 10 more protons per million parallel to the field than anti-parallel. But since there are lots of protons, the magnetization is sufficient. Note that this effect also occurs in the presence of the Earth's magnetic field, but since this is many orders of magnitude weaker it is more challenging to detect.

5.4 Transverse Magnetization

So far, we have only considered a static external magnetic field, B_0, oriented in the z-direction, which causes the spins to precess about the z-axis at the Larmor frequency. This results in a net magnetization $\mathbf{M}$ that is parallel to $\mathbf{B}_0$, with no transverse component, and a longitudinal component M_0.

Because $\mathbf{M}$ is a magnetic moment, it experiences a torque if an external, time varying, magnetic field $\mathbf{B}(t)$ is applied (recall $\mathbf{J}$ is the bulk angular momentum):

$$\frac{d\mathbf{J}(t)}{dt} = \mathbf{M}(t) \times \mathbf{B}(t) \tag{5.7}$$

$$\frac{d\mathbf{M}(t)}{dt} = \gamma \mathbf{M}(t) \times \mathbf{B}(t) \tag{5.8}$$

where to get Eq. (5.8) we have assumed we are only interested in a 'short' period of time and used Eq. (5.4).

If $\mathbf{B}(t) = B_0 \hat{\mathbf{z}}$ and $\mathbf{M}$ happens to be oriented at an angle α with respect to the z-axis with initial value M_0, then the solution to this equation as a function of time using polar coordinates, as in Fig. 5.3, is:

$$M_x(t) = M_0 \sin \alpha \cos(-\omega_0 t + \phi) \tag{5.9}$$

Fig. 5.3 Magnetization system described using polar coordinates

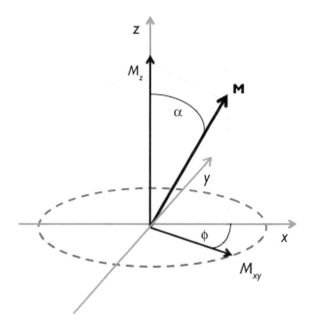

$$M_y(t) = M_0 \sin\alpha \sin(-\omega_0 t + \phi) \tag{5.10}$$

$$M_z(t) = M_0 \cos\alpha \tag{5.11}$$

Exercise 5B
By substitution, show that Eqs. (5.9)–(5.11) are a valid solution to Eq. (5.8) and satisfy the initial condition that the magnetization is oriented at an angle α to $\hat{z}$, with arbitrary orientation in the x–y plane. Hence, determine the value for the precession frequency ω_0.

The frequency of precession, as you show in Exercise 5B, is just the Larmor frequency (Eq. 5.2) for the proton (or whichever nucleus is being used) in the static field. As we might expect, when $\alpha = 0$ there is no transverse magnetization ($M_x = M_y = 0$) and $M_z = M_0$. When $\alpha \neq 0$, then the equations tell us that the *net magnetization* is precessing around $\mathbf{B}_0$ at the Larmor frequency (remember before it was only the individual protons that were precessing). Note that we have not worried yet about how we achieved the condition that $\mathbf{M}$ was oriented at an angle to the z-axis ($\alpha \neq 0$).

We can decouple the longitudinal component $M_z(t)$, which is oriented along the axis of the external magnetic field $\mathbf{B}_0$, from the transverse component $M_{xy}(t)$, which lies in the plane orthogonal to the direction of $\mathbf{B}_0$, as has been done in Fig. 5.3. $M_{xy}(t)$ can be written as a complex number:

$$\mathbf{M}_{xy}(t) = \mathbf{M}_x(t) + \mathrm{j}\mathbf{M}_y(t) \tag{5.12}$$

with phase angle $\phi = \tan^{-1}\left(\frac{M_y}{M_x}\right)$. The transverse magnetization then becomes:

$$\mathbf{M}_{xy}(t) = M_0 \sin\alpha\, e^{-\mathrm{j}(\omega_0 t - \phi)} \tag{5.13}$$

Exercise 5C
Show that Eq. (5.13) is the same as Eqs. (5.9) and (5.10) if we interpret the real part of $\mathbf{M}_{xy}$ as the x-component of $\mathbf{M}$, and the imaginary part as the y-component.

This transverse magnetization is rapidly rotating and thus creates a radiofrequency (RF) excitation, which in turn will induce a voltage (which is a measurable signal) in a coil of wire outside the sample. As you can probably work out, if you want to maximize the transverse magnetization, and thus the size of the measurable signal in a coil of wire, we want $\alpha = 90°$.

In Exercise 5A, you found that the Larmor frequencies for typical clinical MRI systems were close to the RF frequencies used to broadcast FM radio stations. What we have found in this section is that the signals generated using NMR will be RF radiation at the Larmor frequency. Inevitably, these signals will be small, remember we are reliant on only a tiny fractional difference between protons in the up and down states. This means we are likely to see interference between stray radio signals from outside the MRI system. This is why MRI scanners are always found in rooms that incorporate a Faraday cage into the walls to keep out any inferring RF. Evidence for this can generally be found in the window between the scan room and the control room, in the form of a very fine mesh in the glass.

5.5   RF Excitation

At equilibrium, $\mathbf{M}(t)$ is fully aligned with the static field $\mathbf{B}_0$, i.e. $\alpha = 0$. So far, we have ignored how we can get the system into a state in which $\alpha \neq 0$. Let us now impose a very small magnetic field $\mathbf{B}_1$, oriented in some specific orthogonal direction to $\mathbf{B}_0$, e.g. in x. This extra field will add to the main magnetic field. We could predict that $\mathbf{M}(t)$ will now move into the xy-plane in an attempt to reach equilibrium with the new total applied magnetization. However, once it comes out of equilibrium, $\mathbf{M}(t)$ will also start to precess around z. Thus, if we want to be effective in pushing $\mathbf{M}(t)$ we need to track it as it precesses about the z-axis, and apply a $\mathbf{B}_1$ field whose orientation will push it down to the xy-plane. We might think of this like trying to push someone on a swing; if you time your pushes with the periodic motion of the swing, you are much more effective.

The solution to this problem is that we need to apply a $\mathbf{B}_1$ that is rotating at the same frequency at which $\mathbf{M}$ is precessing about z, which will be the Larmor frequency. Whenever the precessing vector $\mathbf{M}$ coincides with the direction of $\mathbf{B}_1$, it is pushed down towards the transverse plane. This is called linearly polarized RF excitation because its $\mathbf{B}_1$ field is only oriented along one linear direction. Alternatively, we could apply another RF field in direction of $\mathbf{B}_1$, but this time using a quadrature (sine instead of cosine) so that we can continuously push down $\mathbf{M}$. This is called circularly polarized RF excitation, and most coils use this approach. The RF field can be modelled as a complex magnetic field in the transverse plane:

$$B_1(t) = B_1^e(t)e^{-j(\omega t - \phi)} \tag{5.14}$$

where $B_1^e(t)$ is the envelope of the B_1 field amplitude and ϕ its initial phase. The actual evolution of $\mathbf{M}(t)$ now becomes a spiral from the z-axis towards the xy (transverse) plane in a clockwise orientation.

The final flip angle depends on the amplitude and duration of $B_1^e(t)$:

$$\alpha = \gamma \int_0^{\tau_p} B_1^e(t) dt \tag{5.15}$$

where the RF has been applied for duration τ_p. If we reach the transverse plane, it is a 90° pulse, which produces the maximum signal. A flip of 180° is called an inversion pulse as it inverts $\mathbf{M}$ to be aligned with $-\hat{\mathbf{z}}$. For a short RF burst with a constant B_1 amplitude, we can approximate this equation to:

$$\alpha = \gamma B_1 \tau_p \tag{5.16}$$

5.6 The Rotating Frame

It is sometimes more convenient to describe the evolution of $\mathbf{M}(t)$ in a 'rotating frame': one in which we choose axes that are rotating at the Larmor frequency. We can relate the stationary and rotating coordinates by:

$$x' = x\cos\omega_0 t - y\sin\omega_0 t \tag{5.17}$$

$$y' = x\sin\omega_0 t + y\cos\omega_0 t \tag{5.18}$$

$$z' = z \tag{5.19}$$

This allows us to write the transverse magnetization as

$$\mathbf{M}_{x'y'}(t) = M_o\sin\alpha e^{j\phi} = M_{xy}e^{j\phi} \tag{5.20}$$

Comparing this to Eq. (5.13), you will see we have managed to remove the dependence on time (it is our axes that now rotate with time, at a fixed rate ω_0) and that this vector is stationary in our new frame of reference. Note that here we have defined $M_{xy} = \sin\alpha$: the magnitude of the magnetization vector. When we measure our signal, we get both a magnitude and phase values, although we tend to discard the phase information.

Exercise 5D
Confirm that under the transformation in Eqs. (5.17) and (5.18), $\mathbf{M}_{xy}$ can be rewritten in the rotating coordinate system $x'y'$ as in Eq. (5.20).

Figure 5.4 illustrates what happens to the magnetization when we apply our excitation pulse as viewed in our rotating frame of reference. With the rotating component 'removed', it appears as if the vector just tips away from z, towards the xy-plane.

Fig. 5.4 Process of applying an RF excitation pulse to the magnetization as viewed in the rotating frame of reference. The magnetization appears to tip during the application of the RF pulse from alignment with B_0 towards the xy-plane. Reproduced and modified with permission from Short Introduction to MRI Physics for Neuroimaging (www. neuroimagingprimers.org)

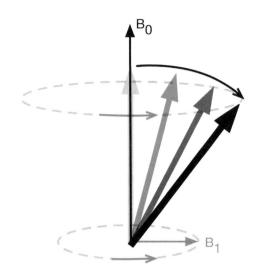

5.7 Relaxation

Whilst applying a B_1 pulse, **M** will precess in a spiral from the z-axis towards the xy (transverse) plane. It will now continue to precess in the new 'orientation'. In our stationary frame, this is the same as tipping M from the z' axis down towards the $x'y'$ plane, and then, it stays there. In reality, this motion is damped, subject to two relaxation processes: transverse relaxation and longitudinal relaxation.

5.7.1 Transverse Relaxation: T_2

Transverse relaxation is also known as spin–spin relaxation. The random microscopic motion of the spins is perturbed by the magnetic moments of neighbouring spins, leading to spins momentarily slowing down or becoming faster. This causes them to dephase relative to one another.

  Another way to think of this is that nearby protons affect the magnetic field that any given proton experiences and this subtly alters its Larmor frequency. Thus, it might precess slightly faster or slower than other protons nearby. Over time, the phase difference between individual protons will increase, as shown in Fig. 5.5. When we sum all these protons together (in our acquired signal) this has the effect of reducing the overall signal as the individual contributions from all the protons are no longer aligned perfectly.

  If we monitor the resulting signal using a receive coil, we will see something like a decaying sinusoid that is called the free induction decay (FID). The signal decay caused by transverse relaxation can be modelled as an exponential decay, with time constant $T2$. The magnitude of the transverse magnetization being described by:

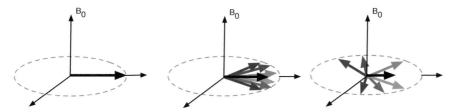

Fig. 5.5 Process of $T2$ relaxation shown in the rotating frame after a 90° flip angle. Over time, the individual contributions from different protons get out of phase with each other. Reproduced with permission from Short Introduction to MRI Physics for Neuroimaging (www.neuroimagingprimers.org)

$$M_{xy}(t) = M_o \sin\alpha e^{-\frac{t}{T_2}} \tag{5.21}$$

5.7.2 Longitudinal Relaxation: T_1

Longitudinal relaxation is also known as spin–lattice Relaxation. It concerns the longitudinal component of the magnetization, $M_z(t)$, as it recovers back to the equilibrium value of M_0. This is modelled as a rising exponential, which also leads to a loss of NMR signal. The longitudinal magnetization is given by a classic exponential recovery expression:

$$M_z(t) = M_0 \cos\alpha + (M_0 - M_0 \cos\alpha)\left(1 - e^{-\frac{t}{T_1}}\right) \tag{5.22}$$

Note that $M_0 \cos\alpha$ is the longitudinal component of the magnetization that was a result of the α-pulse that was applied. We can rearrange this equation to give the more usual form:

$$M_z(t) = M_0\left(1 - e^{-\frac{t}{T_1}}\right) + M_0 \cos\alpha e^{-\frac{t}{T_1}} \tag{5.23}$$

After an excitation pulse, we get both T_1 recovery and T_2 decay happening simultaneously. But, note that they are independent processes, it is not simply a matter of the magnetization tipping backup towards the z-axis again following the path that the excitation drove it down. Typically, T_2 decay is faster than T_1 recovery, so the path that the magnetization takes as it recovers will look a bit like Fig. 5.6. For tissues in the human body, we typically have:

$$250 \, \text{ms} < T_1 < 2500 \, \text{ms}$$

$$25 \, \text{ms} < T_2 < 250 \, \text{ms}$$

Fig. 5.6 Reduction in the
magnetization due to T_2
decay that is more rapid than
T_1 recovery. Reproduced
with permission from Short
Introduction to MRI Physics
for Neuroimaging (www.
neuroimagingprimers.org)

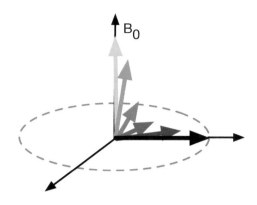

Note that now we have a definition of typical timescales in NMR, and we can define what we meant by 'long' in Sect. 5.3: we need to have left enough time for the recovery according to both T_1 and T_2 to have occurred. We will consider T_1 and T_2 further in Chap. 11 where we will explore how we can use these two different processes to generate contrast between different tissues and thus not just have to rely on differences in proton density.

Exercise 5E

Sketch typical recovery curves for the longitudinal and transversal magnetization after applying a 90° pulse when $T_2 \ll T_1$.

5.8 The Bloch Equations

The Bloch equations pull together the three important properties of the spin system that we have met: proton density, longitudinal relaxation and transverse relaxation. They model the combination of the forced and the relaxation behaviour of the spin system in matrix form:

$$\frac{d\mathbf{M}}{dt} = \gamma \mathbf{M}(t) \times \mathbf{B}(t) - \mathbf{R}(\mathbf{M}(t) - \mathbf{M}_0) \tag{5.24}$$

where $\mathbf{B}(t)$ is the composed of the static and time-varying RF field,

$$\mathbf{B}(t) = \mathbf{B}_0 + \mathbf{B}_1(t) \tag{5.25}$$

and **R** is the relaxation matrix:

$$\mathbf{R} = \begin{pmatrix} 1/T_2 & 0 & 0 \\ 0 & 1/T_2 & 0 \\ 0 & 0 & 1/T_1 \end{pmatrix} \tag{5.26}$$

5.9 Spin Echoes and T_2*

The T_2 or transverse Relaxation is caused by the dephasing of spins. The true free induction decay observed in practice is characterized by a shorter decay rate, called T_2*. The T_2* is due to different protons being in slightly different *chemical environment* and thus having subtly different precession rates, thus, introducing further dephasing. This effect is not a 'true' relaxation effect as it is not random, unlike T_2. T_2* usually dominates over T_2 decay, but we can do something about the T_2* effect. We can recover the T_2* dephasing by forming a spin echo, as illustrated in Fig. 5.7. After the initial 90° the protons start to dephase, faster precessing ones get ahead and slower ones lag behind. If we then apply a refocusing (180°) pulse around the y-axis, the faster spins will now be 'behind' and start to catch up and likewise the slower ones fall back. Thus, after an equal period of time everything should be back in phase again. The spin echo can only recover the additional decay that gives rise to T_2*, but never recovers the (random) T_2 decay itself.

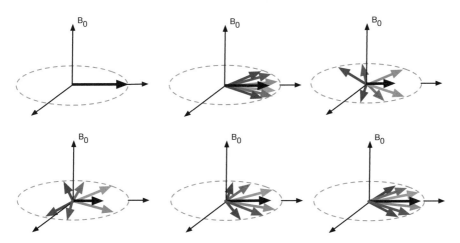

Fig. 5.7 Formation of a spin echo. Top row: spins start to dephase due to different rates of precession after they have been flipped into the transverse plane. A 180° inversion pulse is applied then the spins start to come back into phase, but perfect recovery of signal is not achieved due to T_2 effects. Reproduced with permission from Short Introduction to MRI Physics for Neuroimaging (www. neuroimagingprimers.org)

Further Reading

For further information on nuclear magnetic resonance for imaging see:

Introduction to Medical Imaging, Nadine Barrie Smith and Andrew Webb, Cambridge University Press, 2011, Chapter 5.

Introduction to Functional Magnetic Resonance Imaging, 2nd Ed, Richard B Buxton, Cambridge University Press, 2009, Chapters 3 and 6 (and the Appendix).

Medical Imaging: Signals and Systems, 2nd Ed, Jerry L Prince and Johnathan M Links, Pearson, 2015, Chapter 12.

Webb's Physics of Medical Imaging, 2nd Ed, M A Flower (Ed), CRC Press, 2012, Chapter 7.

Robert W. Brown, Y.-C. Norman Cheng, E. Mark Haacke, Michael R. Thompson, and Ramesh Venkatesan, Magnetic Resonance Imaging: Physical Properties and Sequence Design, 2nd Ed, Wiley-Blackwell, 2014.

Part II
To Images …

Chapter 6
Imaging

Abstract In this chapter, we will start to consider how signals obtained from medical imaging systems are converted into images. We will define some key image-related concepts and then consider some general principles including image restoration, frequency analysis and sampling.

So far, we have considered how to generate signals that contain information about the internal structure of the body. Our task now is to consider how these signals can be turned into images. Namely, how we can resolve these signals to give us specific information about locations in space within the body that we can represent in the form of images. Importantly, our images are in the most part going to be full three-dimensional volumes, although often we only view a single two-dimensional picture at any given time. Whilst we are going to be concerned with assembling 3D information, for our purposes, it is often going to be easier to consider the equivalent 2D problem and assume that extending it to 3D is straightforward. In general, the signals we have generated are going to contain a mixture of the spatial information that we are interested in and thus we are going to have to apply some (mathematical) process to reconstruct the image.

We are going to consider image reconstruction under three categories:

- Timing-based
- Back-projection
- Fourier.

This will allow us to cover the main approaches to converting signals to images as used by the different methods we met for generating signals in the first part of this book. We will also consider iterative methods that are now routinely being used in a wide range of reconstruction problems and increasingly taking over from more 'traditional' methods. Before that, we will start by defining some common concepts that apply to all imaging methods. An important tool we will need to understand image reconstruction is the Fourier transform, particularly when this is applied to 2D (or even 3D) functions. Thus, this chapter includes a brief refresher to the Fourier transform and an introduction to its application to 2D problems, which you can skip over if you do not need it.

© Springer Nature Switzerland AG 2019   55
M. Chappell, *Principles of Medical Imaging for Engineers*,
https://doi.org/10.1007/978-3-030-30511-6_6

6.1   Resolution

An important feature of any imaging system is the resolution we can achieve—related to what is the smallest object that can be observed.

6.1.1   Pixels and Voxels

For pictures and videos, in 2D, we typically divide our images into discrete elements called pixels and thus consider the number of pixels we have, e.g. 'megapixels' for a CCD device in a camera, and we might also consider the size or density of these pixels, e.g. the higher density available on a 'retina display'. Since we are largely going to be working with 3D data in imaging, we need a volume-based resolution element—a 3D pixel—which we will call a voxel, Fig. 6.1. Thus, we will care about how many voxels our imaging equipment can acquire, normally in x, y and z dimensions. This is effectively the spatial sampling of the image and is often something like $32 \times 32 \times 32$ or $64 \times 64 \times 64$—typically called the matrix size. We will also care about the size of the voxel that we can collect, typically measured in milimetres. Note that voxels do not have to have the same size in every dimension, but when they do, we can say they are isotropic. Likewise, we might choose to collect fewer voxels in a particular dimension.

If we multiply the number of voxels in any given dimension by the size of the voxel in that dimension, we arrive at the dimensions of the region that we are imaging: called the field of view (FOV). This is clearly an important measurement since it defines how much of the body we can get in our image.

Fig. 6.1 A voxel is a single 3D element within the larger 3D image, equivalent to a pixel in a 2D image. This figure is attributed to M. W. Toews Vossman and is used, without modification, from Wikimedia under licence: https://creativecommons.org/licenses/by-sa/2.5

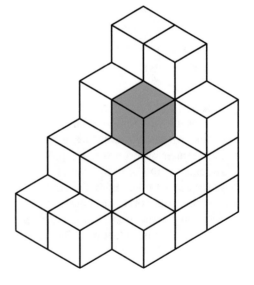

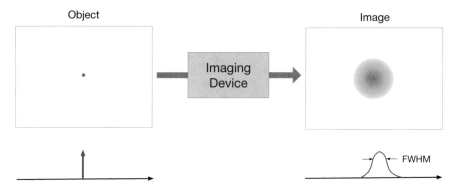

Fig. 6.2 The point spread function (PSF) captures how the imaging device imperfectly reproduces the true object in the image in terms of the effect on an idealized point object

Exercise 6A

An imaging system has a matrix size of $64 \times 64 \times 28$ and is being used to image a FOV of $228 \times 228 \times 140$ mm, calculate the voxel dimensions.

6.1.2 Line and Point Spread Functions

Although we might divide our medical images into little volume elements that we will call voxels, this is not necessarily the 'true' resolution of the data. The voxels are something of an artificial division of the images reflecting the fact that we tend to capture and then reconstruct them digitally. The original information came in the form of a signal and the image device will have certain limitations on how small an object it can capture accurately. This brings us to the concept of the point spread function (PSF), which is the function that describes how an ideal point object (think 3D version of the delta function) appears on the final image, Fig. 6.2. The PSF thus reflects the blurring of an ideal object as captured by the imaging device and can be used to determine the ability of the device to resolve separate objects. There is an analogous concept of the line spread function for imaging methods (e.g. traditional planar X-ray images) that only capture 2D information.

6.2 The General Image Formation Problem

Consider the general image formation problem where our imaging device captures a representation of the object $f(\alpha, \beta)$ as the image $g(x, y)$, Fig. 6.3. Note that we still restrict ourselves to the 2D problem and assume it would be similar in 3D. Since the imaging device is imperfect, information from a single point in f might appear in different places in g. We might write this contribution in terms of a general nonlinear function h:

$$g'(x, y) = h\big(x, y, \alpha', \beta', f'(\alpha', \beta')\big) \tag{6.1}$$

Note that this includes the fact that the intensity of the object itself, i.e. f, will affect the representation in the image. If we can assume that the system behaves (approximately) linearly then:

$$g'(x, y) = h\big(x, y, \alpha', \beta'\big) f'(\alpha', \beta') \tag{6.2}$$

This gives the contribution to x and y in our image from *one* location (α', β') in the real object. Thus, if we want the total intensity, g, we need to perform an integration over the contributions from all locations in the real object:

$$g(x, y) = \int \int h(x, y, \alpha, \beta) f(\alpha, \beta) \mathrm{d}\alpha \mathrm{d}\beta \tag{6.3}$$

If we also assume spatial invariance, then this becomes:

$$g(x, y) = \int \int h(x - \alpha, y - \beta) f(\alpha, \beta) \mathrm{d}\alpha \mathrm{d}\beta \tag{6.4}$$

This is a 2D convolution:

$$g(x, y) = h(x, y) \otimes f(x, y) \tag{6.5}$$

We can identify h as the point spread function, since if f is an ideal point object then $g(x, y)$ is just an image of the blurring effect of the device. The mathematics of the PSF is then just the same as the blurring filter you might apply to a picture in a photograph-editing application.

We could also note that in the Fourier domain:

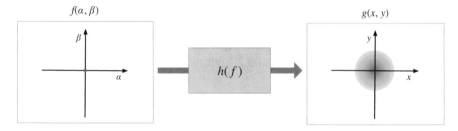

Fig. 6.3 The general image formation problem, the imaging device h captures a representation g of the true object f

$$G(u, v) = H(u, v)F(u, v) \tag{6.6}$$

i.e. the effect of the PSF in space is the same as a multiplication of the frequency domain by H: the **modulation transfer function**.

The actual form of the PSF will vary from device to device and may or may not be the same in all spatial directions. In the very simplest case, we can assume that the PSF is separable, i.e. independent in x and y (and z) thus:

$$h(x, y) = h(x)h(y) \tag{6.7}$$

A widely used form for the PSF is that of a Gaussian distribution, e.g.

$$h(x) = \frac{1}{\sqrt{2\pi\sigma^2}} e^{-x^2/2\sigma^2} \tag{6.8}$$

With σ being a measure of the 'blur' introduced. Conventionally, it is assumed that an imaging device can adequately separate two objects that are farther apart than the full-width half-maximum of the distribution.

Exercise 6B

(a) Show that for the Gaussian PSF in Eq. (6.8)

$$\text{FWHM} = 2\sqrt{2\ln2}\sigma = 2.355\sigma.$$

(b) Hence, for the system in Exercise 6A, if the PSF were to be Gaussian with $\sigma = 2$ mm what would be the 'true' resolution?

6.3 The General Image Restoration Problem

In Sect. 6.2, we have defined the relationship between the object and the image, incorporating the fact that the image will be an imperfect representation of the object. In this section, we will consider the inverse problem: if we have the imperfect image can we recover a faithful representation of the real object? This we will call the image restoration problem. It is worth noting that this is different, but related to, the idea of image reconstruction that we will consider in subsequent chapters. We will return to the more general problem of image reconstruction in Chap. 10.

In Eq. (6.3), we defined the relationship between object and image (assuming linearity) as:

$$g(x, y) = \int \int h(x, y, \alpha, \beta) f(\alpha, \beta) d\alpha d\beta \tag{6.9}$$

We might regard this as an operator equation

$$g = H\{f\} \tag{6.10}$$

where H operates on the object f according to the integral in Eq. (6.9). The image restoration problem is thus one of finding the inverse operator Σ

$$\hat{f} = \Sigma\{g\} = \Sigma\{H\{f\}\} \tag{6.11}$$

The challenge for most medical imaging systems is that although Σ may exist, it is likely to be ill-conditioned, which means that a small perturbation in the image g would lead to a larger, and more problematic, error in the restored image $\hat{f}$.

The main source of perturbations in real images is noise; strictly, we should write the image formation equation as

$$g(x, y) = \int \int h(x, y, \alpha, \beta) f(\alpha, \beta) d\alpha d\beta + n(x, y) \tag{6.12}$$

where $n(x, y)$ represents the noise in the image: a random additional value. It is now possible for multiple different values of object intensity to result in the same value of image intensity depending upon the value of the noise. We can rewrite the equation in discrete form as

$$g_i(x, y) = \sum_{j=0}^{N} h_{ij} f_j(x, y) + n_i(x, y) \tag{6.13}$$

where we have treated the image as being composed of discrete elements, i.e. the voxels, $g_i(x, y)$. We have assumed that the real object can be similarly decomposed, the result being that the relationship between a voxel in the image and elements of the object can be written in terms of a weighted summation. The discretization of the real object is artificial, since it is continuous, but in practice, we work with a discrete image and thus the restored object will also be discrete. If we stack all of the voxels in the image into a vector, we can write an equivalent matrix-vector equation

$$\mathbf{g} = \mathbf{Hf} + \mathbf{n} \tag{6.14}$$

Exercise 6C explores how an ill-conditioned image formation matrix $\mathbf{H}$ gives rise to noise amplification when performing image restoration.

As we saw in Eq. (6.3), the image formation process can be described in terms of a convolution, thus the inverse restoration process is often called deconvolution. The analogy in image processing is of applying a 'sharpening filter', if you do this you will notice that features (edges) in the image get sharper—better defined—but at the same time, noise in the image also becomes more noticeable. Notably, the form of H that is required to implement the discrete version of a convolution with a PSF is poorly conditioned, and hence we would expect deconvolution to degrade the SNR.

Exercise 6C

Starting with the discrete expression for the image formation in Eq. (6.13)

(a) Use matrix algebra show that the following is an expression for the reconstructed object

$$\hat{\mathbf{f}} = \mathbf{f} + \mathbf{H}^{-1}\mathbf{n}$$

(b) Hence, show that the relationship between the SNR of the image and that of the restored object is given by

$$\text{SNR}_{\text{ob}} \leq \frac{\text{SNR}_{\text{im}}}{\left| \mathbf{H}^{-1} \right|}$$

where SNR is defined in terms of the Euclidean norm (2-norm), e.g.

$$\text{SNR}_{\text{im}} = \frac{|\mathbf{f}|}{|\mathbf{n}|}$$

How does this result illustrate that having an ill-conditioned restoration operation results in noise amplification?

(c) By minimizing $|\mathbf{n}| = \mathbf{n}^{\mathrm{T}}\mathbf{n}$, show that the solution in part (a) is the one that minimizes the squared error on $\hat{\mathbf{f}}$.

(d) If instead the squared error on $\hat{\mathbf{f}}$ we define a cost function

$$\tau |\mathbf{Qf}| + |\mathbf{g} - \mathbf{Hf}|$$

Show that the minimum occurs when

$$\hat{\mathbf{f}} = \left(\tau \mathbf{Q} + \mathbf{H}^{\mathrm{T}}\mathbf{H} \right)^{-1} \mathbf{H}^{\mathrm{T}}\mathbf{g}$$

Consider the role of the matrix $\mathbf{Q}$ in 'constraining' the convolution process, (it may be helpful to consider the simple case where $\mathbf{Q}$ is the identity matrix). How might we interpret the role of $\mathbf{Q}$ in the expression $\mathbf{Qf}$?

6.4 A Revision of Frequency Analysis

In Sect. 6.2, we started to use a Fourier representation of images to define the modulation transfer function. Fourier transforms and the Fourier domain, and more generally considering frequency components, will be very helpful when it comes to thinking about image reconstruction. You will hopefully already be familiar with the idea of the Fourier transform of a signal and the idea that signals contain 'frequency' content. This section will provide a very brief refresher of frequency analysis, for more detailed introduction of the topic see the Further Reading for this chapter.

6.4.1 Fourier in One Dimension

The key assumption we make when doing a Fourier transform is that a function can be represented by an infinite sum of sinusoidal components. The Fourier transform produces a function with both magnitude and phase for each of these components, i.e. it tells us how much of a sinusoidal component at a given frequency we need to include and its phase relative to the other components. The results of the Fourier transform can be described as a frequency spectrum, most often this is just the magnitude part, and we pay less attention to the phase.

If the original function is periodic, then we get only a fixed set of (harmonic) components: a Fourier series. For a general function, we need an integral over all possible frequencies; the Fourier transform:

$$F(u) = \int_{-\infty}^{\infty} f(x) e^{-2\pi j u x} \, \mathrm{d}x \qquad (6.15)$$

Since we are going to be dealing with images with a spatial dimension, rather than time series, this version of the Fourier transform is written in terms of x and the equivalent 'spatial' frequency u, whereas for time series, frequency was the number of cycles per second (of a sinusoidal variation). In this case u is the number of cycles per unit distance. The Fourier transform (and its inverse) allows us to go back and forth between the spatial (time) domain and the frequency, or Fourier, domain representations of a function.

It is worth also noting some key properties of Fourier transforms, including:

- Signals with lots of sharp edges, what we might call 'details', need high frequencies to represent them, i.e. they have a broad frequency spectrum or equivalently a high bandwidth. You will explore this in Exercise 6D.
- **Duality**: the FT of a particular shape of signal is the same as the inverse Fourier transform of the same shaped Fourier spectrum.
- **Convolution**: the convolution of two signals in the space (time) domain is the same as the multiplication of their respective Fourier transforms in the frequency domain.

$$f(x) \otimes g(x) = \int_{-\infty}^{\infty} f(\lambda) g(x - \lambda) \, \mathrm{d}\lambda = F(u)G(u) \qquad (6.16)$$

(We used the 2D version of this property in Eq. [6.6]).

Exercise 6D

Consider two functions defined in terms of a spatial variable x: a rectangular function and a Gaussian function. Find and sketch the Fourier transform of each. Assuming that the functions represent simple 1D 'images', what implications does the form of Fourier transform you have found have on our ability to capture these images if our imaging device has limited bandwidth, i.e. we cannot capture some or all higher frequencies that might exist?

6.4.2 Fourier in Two Dimensions

For imaging, we will need to extend our definition of Fourier transforms to more dimensions. Again, we will constrain ourselves to 2D examples. The FT in 2D is given by:

$$F(u, v) = \iint\limits_{-\infty}^{\infty} f(x, y) e^{-2\pi j(ux+vy)} \mathrm{d}x \mathrm{d}y \qquad (6.17)$$

The inverse being given by:

$$f(x, y) = \iint\limits_{-\infty}^{\infty} f(u, v) e^{2\pi j(ux+vy)} \mathrm{d}u \mathrm{d}v \qquad (6.18)$$

We now have two Fourier domain frequency components: u and v. This means that if we want to plot the magnitude (or phase) of the Fourier transform, we will need to visualize it as a surface or as an image. The meanings of the frequency components of the 2D FT are similar to the 1D case, i.e. we want to build our signal/image from the sum of sinusoidal components and the magnitude of the FT tells us magnitude of each of the different components we need. The difference with the 2D case is that we now have components with sinusoidal variation in magnitude in a specific direction, and some examples are given in Fig. 6.4. Just as any signal could be composed of a sum (integral) of Fourier components with different frequencies, now any image can be composed of a sum (integral) of components with different frequencies and directions.

Just like in the 1D case, we get both magnitude and phase information with our Fourier transform, but we often ignore the phase. We can plot the 2D Fourier magnitude information and thus examine the 2D frequency content of the image. An example is shown in Fig. 6.5. Each location in the magnitude plot corresponds to a specific spatial frequency and direction (i.e. the magnitude of a Fourier component

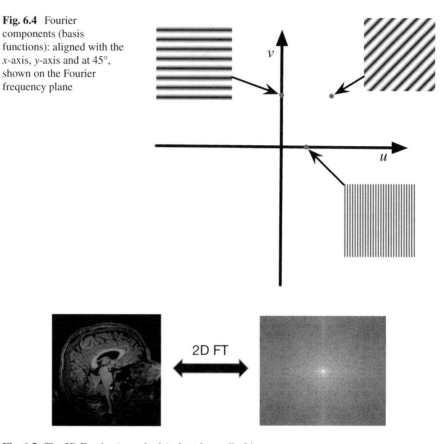

Fig. 6.4 Fourier components (basis functions): aligned with the x-axis, y-axis and at 45°, shown on the Fourier frequency plane

Fig. 6.5 The 2D Fourier (magnitude) plot of a medical image

like one of those in Fig. 6.4). The axes of this plot are u and v, with values on the u-axis corresponding to sinusoidal variations in x in the original image, and those on v corresponding to sinusoidal components in y, as shown in Fig. 6.4. We tend to put the origin at the centre, like 1D Fourier we have both positive and negative frequency components. Values closer to the centre thus correspond to slow variations in intensity in the image and those further out to high frequencies, thus the fine details are captured by the outer edges of the Fourier/frequency space.

6.4.3 2D FT in Polar Coordinates

It is also going to be useful to define the 2D Fourier transform in polar coordinates:

$$F(\rho, \phi) = \int\limits_{0}^{\pi} \int\limits_{-\infty}^{\infty} f(r, \theta) e^{-2\pi j \rho r \cos(\phi - \theta)} |r| dr d\theta \qquad (6.19)$$

Likewise, the inverse FT:

$$f(r, \theta) = \int\limits_{0}^{\pi} \int\limits_{-\infty}^{\infty} F(\rho, \phi) e^{2\pi j \rho r \cos(\phi - \theta)} |\rho| dr d\theta \qquad (6.20)$$

Just as u and v are the Cartesian coordinates of our 2D Fourier space, ρ and φ are the polar coordinates in frequency space, i.e. they can be used to specify a location in the Fourier magnitude plot (such as that in Fig. 6.5) in terms of a distance from the origin and angle to the u-axis.

6.5 Sampling

We have already assumed in this chapter that the image we acquire exists in terms of discrete values associated with volumes in the real object: the voxels. Since the real object is continuous, this implies that our image is a sampled version of the true object, and thus we need to understand the implications of sampling on the way in which the image is a faithful representation of the real object.

Sampling is the process of converting a continuous time signal to an equivalent discrete representation by recording the value of the continuous signal at (typically) regular time instants. Mathematically, this can be represented as the multiplication of the signal with a sampling function, which is a train of evenly spaced delta functions (sometimes called a comb function), whose spacing determines the sampling rate, normally quoted in Hz. We can define the equivalent process for images where we say that the discrete image is a sampled version of the true object, as if it has been multiplied by a sampling function represented by a regular grid of delta functions in 2D or 3D. Thus, the sampling rate for an image is directly related to the dimensions of the voxel. As we noted in Sect. 6.1, we tend to define images in terms of matrix size and voxel dimensions, rather than sampling 'rate'. Unsurprisingly, the smaller the features we wish to visualize in the image the smaller our voxels need to be, which is equivalent to using a 'faster' sampling rate in the spatial dimension of the image.

The Nyquist sampling theorem tells us that if we want to be able to faithfully reproduce the original continuous signal from the sampled version, we need to sample at twice the highest frequency component in the original signal. Otherwise, we will experience aliasing, where frequencies present above the Nyquist limit, i.e. half the sampling frequency, will appear as spurious extra low-frequency components in the reproduced signal. We can understand this process via Fourier theory by noting that the signal has a frequency spectrum that will extend only as far as the highest frequency in the signal, i.e. it is bandlimited. If we sample the signal, this has the effect of replicating the frequency spectrum in the frequency domain with spacing that is determined by the sampling rate: a higher sampling rate gives rise to a wider spacing of replicas in the frequency domain.

The effect of under-sampling and aliasing is illustrated in Fig. 6.6, and it can be understood in terms of sampling being represented as a multiplication of the original signal with a train of delta functions, which gives rise to a convolution in the frequency domain of the spectrum of the signal with the spectrum of the sampling functions, itself a train of delta functions. Thus, in the frequency domain, the sampled signal is represented by repetitions of the frequency spectrum of the original signal spaced at the sampling frequency. If the sampling rate is not sufficiently rapid, i.e. the case of under-sampling, the result in the frequency domain is an overlap of the replicated spectra. At the point where the sampled signal is reconstructed into a continuous signal, extra spurious frequency components appear.

The effect of aliasing in images is that it causes artefacts when the sampling, i.e. the voxel size, is insufficient to capture the high frequencies, i.e. fine details, in the true object, and this is illustrated in Fig. 6.7. Since in practice we cannot achieve an arbitrarily small voxel size due to acquisition constraints, some form of anti-aliasing filtering will be required.

It is worth noting that there is another, related but separate, sampling process that takes place in an imaging system at the point where we capture the continuous signals that we met in Chaps. 2 through 5. This is necessary because imaging systems are digital and thus detection will involve an analogue-to-digital convertor. This sampling process does not very accurately approximate the ideal grid of delta functions we have assumed can represent the process of sampling. In practice, the detection process actually captures an integral of the signal over a region, and this might be an area

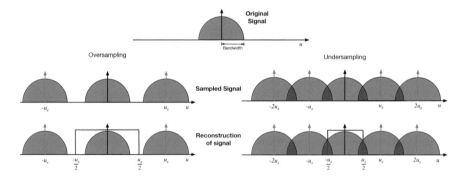

Fig. 6.6 An illustration of the process of sampling viewed in the frequency domain. The original signal is composed of a range of frequencies as shown in the frequency spectrum; this example has a limited bandwidth. The effect of sampling in the time domain is to reproduce the spectrum in the frequency domain at a spacing equal to the sampling frequency. Where this is smaller than twice the bandwidth there is an overlap of the replicas, which will sum to give a larger contribution in the spectrum at the overlapping frequencies (summation not shown). The process of reconstructing the signal, i.e. producing a continuous signal from the samples, is, in the ideal case, the equivalent of selecting frequencies in a region defined by the sampling frequency. For the under-sampled case, this leads to spurious extra, aliased, frequencies appearing in the reconstructed signal

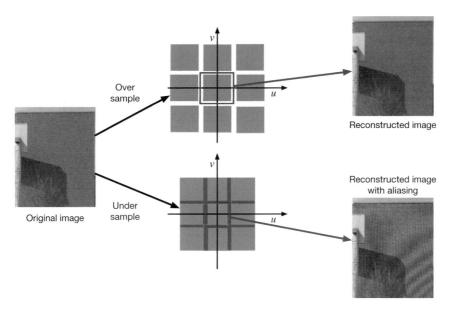

Fig. 6.7 An example of sampling and aliasing effects for images. When under-sampling occurs, frequency contributions overlap in 2D frequency space resulting in aliasing artefacts in the image. In this case, a Moiré effect can be seen in the aliased image. Photographic images reproduced from Wikimedia (user Cburnett), used under the Creative Commons Attribution-Share Alike 3.0 Unported license

of detector, for most of the signals we have met, or arising from a volume for NMR signals. This, in itself, causes some filtering or blurring of the spatial information contained in the signal and thus in most cases achieves the effect of an anti-aliasing filter. It is this process that, at least partially, gives rise to the PSF that we met in Sect. 6.1 and means that the true resolution of the image might not match the resolution defined by the grid of voxels we choose for the image. In fact, to avoid aliasing we need to ensure that the voxel size used is not a lot larger than that defined by the PSF.

Further Reading
For more details of general image characteristics, and image formation, restoration and processing see

Introduction to Medical Imaging, Nadine Barrie Smith and Andrew Webb, Cambridge University Press, 2011, Chapter 1.
Medical Imaging: Signals and Systems, 2nd Ed, Jerry L Prince and Johnathan M Links, Pearson, 2015, Chapters 2 and 3.
Webb's Physics of Medical Imaging, 2nd Ed, M A Flower (Ed), CRC Press, 2012, Chapter 11.

Introductions to the Fourier transform (and series) can be found in many undergraduate mathematics texts. Some fairly comprehensive introductions in the context of
medical imaging can be found in

Introduction to Medical Imaging, Nadine Barrie Smith and Andrew Webb, Cambridge University Press, 2011, Chapter 1.

Medical Imaging: Signals and Systems, 2nd Ed, Jerry L Prince and Johnathan M
Links, Pearson, 2015, Chapter 2.

Chapter 7
Timing-Based Reconstruction—Ultrasound

Abstract In this chapter we will look at timing-based reconstruction, taking ultrasound imaging as our example. We will see how very simple timing information can be used to reconstruct images and go on to consider how we can focus ultrasound and steer the ultrasound beam to make 2D and 3D imaging feasible.

Timing-based represents the simplest reconstruction problem we will consider. As the name suggests, it relies on there being timing information in the signal that relates to position within the body, and this means it generally applies to reflection-based methods. In Chap. 3, we saw that ultrasound is based on recording the reflected waves from interfaces in the body from ultrasound waves that we have generated in our transducer. For imaging, we rely upon pulses of ultrasound, launching a pulse and then 'listening' to the echoes that come back. Since the speed of sound is finite, reflected waves from interfaces that are deeper in the body will take longer to be received than those from nearer interfaces, providing a direct way to infer depth from the received signal. Whilst ultrasound can be either continuous wave or pulsed, you can appreciate that using continuous wave, ultrasound would present a challenge if we want to localize the reflections. It is useful if we only want Doppler measurements without localization.

7.1 Converting Time to Depth

Since the timing of received energy from reflections of the ultrasound pulse is directly related to distance, if we are prepared to assume a uniform speed of sound, we can immediately generate an 'image' along the line of the ultrasound beam; this would be called **A-Mode imaging**. This form of imaging is shown conceptually in Fig. 7.1. Here, we have launched a pulse into the body and received a series of echoes from the various interfaces A, B and C. The first to arrive represents the shallowest interface and thus a plot of received intensity as a function of time can be directly translated into a one-dimensional 'image'. A-mode imaging is not quite the sort of image we

© Springer Nature Switzerland AG 2019
M. Chappell, *Principles of Medical Imaging for Engineers*,
https://doi.org/10.1007/978-3-030-30511-6_7

Fig. 7.1 Process of creating an A-mode ultrasound image from reflections received from interfaces within an object

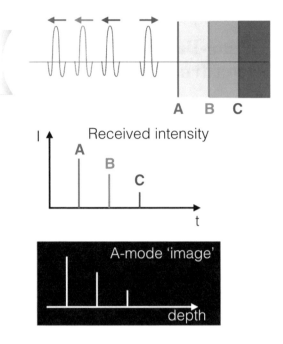

might hope to achieve, it is only 1D, and is not all that common in use, but it can still be useful in specific contexts.

This simple way of forming an A-mode image neglects the effect of attenuation. The result is that, whilst interfaces that are deeper in the body can be identified, the amplitude of the echo will be smaller, as indicated in Fig. 7.1. To correct for this, we use time gain compensation (TGC) which attempts to amplify pulses in proportion to the depth from which they originate. To do this, we could multiply the measured signal amplitude by the inverse of the equation we used to model attenuation in Chap. 3. This requires the information we have gained on the depth of the interface from the timing of the echoes and the assumption of a uniform speed of sound. We also need to specify the relevant attenuation coefficient; again, the simplest correction assumes a uniform value. In practice, this is an oversimplification of the TGC needed for a pulse-echo ultrasound system taking into account transmission and reception by the transducer.

Exercise 7A You are operating an ultrasound imaging device using a 2.5 MHz transducer. You seek to image deeply in breast parenchyma and structural tissue possessing an attenuation coefficient of 0.87 dB cm^{-1} $MHz^{-1.5}$.

(a) Assuming a (standard) uniform speed of sound of 1540 m s^{-1}, determine the depth of the interfaces which give rise to echoes at 1.3 and 6.5 μs.

(b) Time gain compensation (TGC) is required to correct for attenuation, show that the reflected signal that is received at time t (after the pulse was launched) should be multiplied by the factor:

$$\text{TCG}(t) = e^{\mu c t}$$

(c) Hence, determine

   (i) The TGC required for the two echoes in part (b).
   (ii) The TGC required if you increase the ultrasound frequency to 5 MHz?

The axial resolution of an ultrasound image is determined by the pulse duration, p_d, since we need to distinguish between the echoed pulses from two interfaces to be able to resolve them. If we assume that we can distinguish between two pulses that overlap by at most half their duration, then:

$$\text{Axial resolution} = \frac{1}{2} p_d c \qquad (7.1)$$

If we were to repeat our A-mode scan and display each new line beside the next in a scrolling display we would arrive at a M-mode 'image', which can be useful for highly mobile tissues, e.g. in the heart.

7.2 Ultrasound Beams and Pulse-Echoes

The model of ultrasound pulse-echo imaging we have considered in Sect. 7.1 was very simplified. It assumed that we launch a pulse into the tissue and it only travels along an axis perpendicular to the transducer, producing echoes that also only return along that axis. In reality, the sound produced by the transducer will be a beam that will have a spatial profile: further away from the transducer the sound spreading out along a cone. Reflections thus returning to the transducer from a volume of tissue across a range of angles.

It is possible to understand some of the implications of this by building a better (although still simplified) model of the behaviour of a pulse-echo when the transducer is a flat plate, as shown in Fig. 7.2.

We can model the pulse as

$$n(t) = n_e(t) e^{-j(2\pi f_0 t - \phi)} \qquad (7.2)$$

Which is in the form of a 'carrier' at the ultrasound frequency, f_0, and an envelope that represents the shape of the pulse, $n_e(t)$, with an arbitrary phase offset, ϕ. We will assume each point on the face of the transducer to be acting as an acoustic dipole,

Fig. 7.2 Geometry for
analysis of a simple flat
transducer

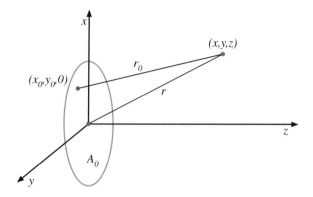

reflecting the fact that the plate moves in the z-direction to create the wave. This
allows us to write the contribution from a point on the face of the transducer to the
acoustic signal at a point as

$$p(x, y, z, t) = \frac{z}{r_0^2} n(t - c^{-1} r_0) \tag{7.3}$$

using the geometry in Fig. 7.2. The total sound arriving at any point in space will be
a superposition of all of the contributions from the surface of the transducer

$$p(x, y, z, t) = \iint_{A_0} \frac{z}{r_0^2} n(t - c^{-1} r_0) dx_0 dy_0 \tag{7.4}$$

where we are integrating over the surface of the transducer A_0. If we describe the
scattered intensity returned from a scatterer located at this point in space in terms of
a function $R(x, y, z)$, then the pressure contribution from scattering at a point on the
face of the transducer $(x_0', y_0', 0)$ is

$$p_s(x_0', y_0', t) = \iint_{A_0} R(x, y, z) \frac{z}{r_0'^2} p(x, y, z, t - c^{-1} r_0') dx_0' dy_0' \tag{7.5}$$

where the dipole pattern of the individual element of the transducer surface has been
taken into account. Thus,

$$p_s(x_0', y_0', t) = \iint_{A_0} \iint_{A_0} R(x, y, z) \frac{z}{r_0'^2} \frac{z}{r_0^2} n(t - c^{-1} r_0 - c^{-1} r_0') dx_0 dy_0 dx_0' dy_0' \tag{7.6}$$

If we make the approximation that the envelope of the pulse arrives at all points
in a given z-plane simultaneously (the plane wave approximation), then

$$n(t) \approx n_e\left(t - 2c^{-1}z\right)e^{-j\left(2\pi f_0\left(t-c^{-1}r_0-c^{-1}r_0'\right)-\phi\right)}$$
$$\approx n\left(t - 2c^{-1}z\right)e^{jk(r_0-z)}e^{jk(r_0'-z)} \tag{7.7}$$

Which allows us to write the received intensity from a single scatterer as

$$r(x, y, z, t) = K R(x, y, z)n\left(t - 2c^{-1}z\right)q(x, y, z)^2 \tag{7.8}$$

with

$$q(x, y, z) = \iint_{A_0} \frac{z}{r_0^2} e^{jk(r_0-z)} dx_0 dy_0 \tag{7.9}$$

where $q(x, y, z)$ represents the field pattern for the transducer, and K is a gain factor related to the sensitivity of the transducer and the various receive electronics.

Assuming superposition, the total signal received from a distribution of scatterers will be

$$r(t) = \int_0^\infty \int_{-\infty}^\infty \int_{-\infty}^\infty K R(x, y, z)n\left(t - 2c^{-1}z\right)e^{-2\mu z}q(x, y, z)^2 dx dy dz \tag{7.10}$$

where the effect of attenuation on the round trip that the echo has taken is also included, using the simplification that to account for attenuation it is approximately sufficient only to consider the distance z from the transducer, not the true round-trip distance.

A useful result is to rewrite this in terms of

$$\tilde{q}(x, y, z) = zq(x, y, z) \tag{7.11}$$

i.e. pulling the z-term out of $q(x, y, z)$, and noting that $z = ct/2$, so that we can write

$$r(t) = K' \frac{e^{-\mu ct}}{(ct)^2} \int_0^\infty \int_{-\infty}^\infty \int_{-\infty}^\infty R(x, y, z)n\left(t - 2c^{-1}z\right)\tilde{q}(x, y, z)^2 dx dy dz \tag{7.12}$$

Now, most of the terms that affect the amplitude of the received signal appear outside the integral. As we noted in Sect. 10.1, we need to correct for attenuation of the signal, and we do this using TGC. This more detailed analysis thus suggests that instead of the equation derived in Exercise 7A, we should do a correction using

$$\text{TGC}(t) \propto (ct)^2 e^{\mu ct} \tag{7.13}$$

Assuming that we do a correction of this form, substituting in the model of the pulse, gives

$$r_c(t) = \int\limits_0^\infty \int\limits_{-\infty}^\infty \int\limits_{-\infty}^\infty R(x, y, z) n_e\big(t - 2c^{-1}z\big) e^{j\left(2\pi f_0\left(t - 2c^{-1}z\right) - \phi\right)} \tilde{q}(x, y, z)^2 \mathrm{d}x\mathrm{d}y\mathrm{d}z$$

$$(7.14)$$

Similarly to the pulse that is launched, this received signal is of the form of a carrier at f_0 modulated by an envelope

$$r_c(t) = r_e(t) e^{-j(2\pi f_0 - \phi)} \tag{7.15}$$

The envelope can be extracted using some form of demodulation. Whilst the size of the envelope depends upon $R(x, y, z)$ and hence the strength of reflection from the scatterers as expected, there are other contributions which mean that it is not trivial to relate amplitude to the strength of reflection (and thus mismatch in acoustic impedance).

In reality, the field pattern, as captured by $q(x, y, z)$, makes an important contribution to the resulting images. To achieve a more focused ultrasound beam, rather than using a flat transducer, a curved shape might be used. Commonly, an array of transducers is used to create both more focused and also steerable beams of ultrasound.

7.3 Scanning

Up to this point, the images that we have been able to produce have only been one-dimensional. We can go one step further and produce a cross-sectional slice through our patient by scanning the ultrasound beam back and forth to create a B-mode image. This will be the familiar view if you have ever seen an ultrasound scan, as shown conceptually in Fig. 7.3. In a B-mode image, the intensity of the pixels is the (time gain compensated) strength of the echo, and the radial axis is directly related to the

Fig. 7.3 The FOV for a B-mode ultrasound reflects the acquisition process of sweeping the beam back and forth through an arc

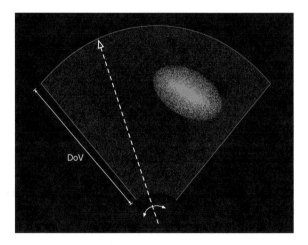

timing of the received echoes and the angular axis arises from the sweeping of the beam back and forth through an arc. This form of scanning, that does not require the array of transducers to be as wide as the FOV we wish to image, gives rise to the unusual shape of the FOV that looks like a segment of a circle. The scanning can be achieved mechanically, but more generally now is done electronically using an array of transducers to create a 'steerable' beam.

Important parameters for B-mode imaging are:

- Depth of view (DoV)—the depth of the deepest interface we can image depends upon how long we wait for echoes to return: the pulse repetition frequency (PRF).
- Scan line density—every time we launch a pulse and listen for the echoes we acquire one line of our image (frame). To give a sufficiently accurate representation, we will need to achieve a given density of scan lines, *lines per frame* (LpF), which in turn will determine, along with the PRF, how long it takes to acquire a full frame.
- Frame rate (FR)—the rate at which scanning takes place will determine how regularly the image is updated, which may be important if we want to track moving tissue. The speed at which we can scan depends upon the pulse repetition frequency and the scan line density.

3D images can be created by assembling a series of B-mode scans and may be called 4D when this is repeated (as often as possible) to update with time. The process of creating a 3D image may be done manually by sweeping the probe along the body and then reassembling all the slices, or by using special probes that scan a full 3D volume rather than just a single slice.

Although we could relatively easily define the axial resolution of an ultrasound image in terms of the pulse duration, it is somewhat more complex to define the lateral resolution. Perhaps unsurprisingly this resolution gets lower further away from the end of the image where the transducer is. The true resolution depends greatly on the focusing of the ultrasound beam and how it is being steered to create the image.

> **Exercise 7B** As in Exercise 7A, you are operating an imaging device using a 2.5 MHz transducer for breast imaging. If the required frame rate is 64 Hz and the depth of view 10 cm, what would be the maximum possible scan line density?

7.4 Artefacts in Ultrasound

There are various processes that give rise to artefacts in ultrasound images. We have already met the concept of speckle in Chap. 3: tissues are full of small structures which are too small and closely packed to be resolved but nevertheless scatter ultrasound.

Interference between the resulting waves gives rise to the familiar appearance of ultrasound images, which look 'noisy' but with a noise that has an inherent spatial pattern to it. Images of certain types of tissue have a characteristic and recognizable 'texture', e.g. liver, kidney and spleen, although this gives no spatial information about the structures themselves.

Speckle noise is a feature of all ultrasound images. There may be further artefacts associated with particular tissues or combinations of tissues. One example is where there are multiple reflections between two highly reflective interfaces and/or an interface and the transducer that appear in the image as extra structures deeper in the FOV. This is reverberation and a good example of when this happens is imaging the diaphragm. This can act as a 'mirror' so that parts of the liver are displayed as if they were in the lung.

In Chap. 3, we considered how highly reflective or attenuating structures, e.g. bone, lung or bowel gas can reduce the amplitude of echoes from regions behind them. The effect on the resulting images is to cast a shadow over the region behind them: hence, such artefacts are called shadowing. Note that the time gain compensation does not correct for this effect; it only provides a correction for approximately uniform attenuation of the ultrasound. More complex intensity reductions due to the interfaces in the imaging region remain in the final image.

When forming the images, we have assumed a uniform speed of sound in the conversion from time to depth. Variability in the true speed will lead to distortion in the individual A-mode scan and thus the final image regions with a faster speed will appear thicker and vice versa. A further effect we noted in Chap. 3 was refraction: a change in direction of the ultrasound beam will occur upon passing through a tissue with very different acoustic impedance, e.g. bone (especially the skull). Ultimately, this will distort the image of the tissue underneath. According to our simple description of A-mode ultrasound, we might expect that distortion would lead only to a loss of signal (and thus an effect like shadowing), since any reflections will not return directly to our probe along the line they were sent. In practice, the transducer has a range of sensitivity (and can be focused), meaning that refracted signals do contribute. A similar distortion effect can arise from dispersion of the ultrasound pulse as it travels through the tissue.

Further Reading

For more information on ultrasonic imaging see

Introduction to Medical Imaging, Nadine Barrie Smith and Andrew Webb, Cambridge University Press, 2011, Chapter 4.
Medical Imaging: Signals and Systems, 2nd Ed, Jerry L Prince and Johnathan M Links, Pearson, 2015, Chapter 11.
Webb's Physics of Medical Imaging, 2nd Ed, M A Flower (Ed), CRC Press, 2012, Chapter 6.

Chapter 8
Back-Projection Reconstruction—CT and PET/SPECT

Abstract In this chapter we will consider the principles of projection and the associated method of back-project reconstruction. We will apply these principles to X-ray Computed Tomography (CT), Single Photon Emission Computed Tomography (SPECT) and Positron Emission Tomography (PET). In doing so, we will explore why PET can achieve a higher resolution than SPECT, but requires a different device design than needed for X-ray CT and SPECT.

A number of the signals we have considered have passed through the body either from some external (X-rays in Chap. 2) or internal source (PET/SPECT in Chap. 4). Thus, the signal we have received contains information about either the tissue along the whole path or concentration of a radiation source within the body. Unlike ultrasound, where we could use timing information in the signal itself to reconstruct information from along the line of the beam, in this case, there is no other information in the signal itself to help us. The problem is fundamentally ill-posed as we need to reconstruct a line of information from a single measurement. We are not totally without hope though. You will already be familiar with a classical planar X-ray image; what you see is a projection through the body. A projection is a 2D representation of a 3D object where the intensity at every point represents the cumulative effect of absorption of the X-rays along their path through the tissue, precisely the idea behind transmission methods in Chap. 2. Inevitably, projection loses the important third dimension of how the tissues vary along the projection line itself.

We can, of course, obtain projections in any direction we wish. So, for example, we might take three orthogonal projections. Quite reasonably, we might then assume that it ought to be possible from these three projections to reconstruct the full 3D volume. In practice, we will use more projections, but this is the essence of back-projection: starting with a series of projections, how do you reconstruct the 3D objection from which they were generated?

© Springer Nature Switzerland AG 2019   77
M. Chappell, *Principles of Medical Imaging for Engineers*,
https://doi.org/10.1007/978-3-030-30511-6_8

8.1   Projection: Computed Tomography

For X-ray-based medical imaging systems, i.e. computed tomography (CT), acquiring projections involves the mounting of source and detector on a gantry and rotating this around the subject to acquire projections from a plane slice. Multiple slices can then be acquired by moving the subject through the gantry. We can write the projection of the object when the gantry is at an angle φ as in Fig. 8.1, as:

$$I_\phi(x') = I_{\phi,0}(x')e^{-\int_{AB}\mu(x,y)dy'} \tag{8.1}$$

where we have defined a 'global' set of axes, xy, that are related to the object/subject and a local set of axes, $x'y'$, aligned with the CT source and detector arrangement—with these sitting along the y' axis (see Fig. 8.1). This equation says that the intensities at any location in the projection, x', depend upon the (attenuated) intensity of the X-rays that would fall on the detector multiplied by the line integral of the attenuation coefficient between the source and detector, along the line AB. To make this work, we have to ensure that we only receive X-rays from the precise projection line between source and detector, requiring the use of the collimators we met in Chap. 2.

If we normalise by the source intensity, we can define the projection as:

$$\lambda_\phi(x') = -\ln\left(\frac{I_\phi(x')}{I_{\phi,0}(x')}\right) = \int_{AB}\mu(x,y)dy' \tag{8.2}$$

which can also be written as

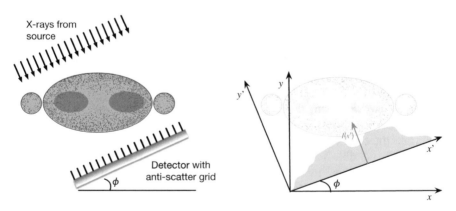

Fig. 8.1   Generation of a single projection for an X-ray system (left), axes system for describing projection (right)

$$\lambda_\phi(x') = \iint\limits_{-\infty}^{\infty} \mu(x, y)\delta(x \cos \phi + y \sin \phi - x')\mathrm{d}x\mathrm{d}y \qquad (8.3)$$

where we are using $\delta(x, y)$, the Dirac delta function, to pick out the path of the line integral in Eq. 8.2 and thus convert into a double integral over space. This is illustrated in Fig. 8.2, and it works because the delta function is zero every where apart from when $x' = x \cos \phi + y \sin \phi$. For any combination of x and y where that is the case, that point is included within the integral. Thus, the integral is computed along a line parallel to the y' axis that corresponds to a specific value of x'.

Exercise 8A A CT system is being used to image the object in Fig. 8.3 which has

Fig. 8.2 Geometry behind the projection equation that allows the line integral to be converted to a 2D surface integral

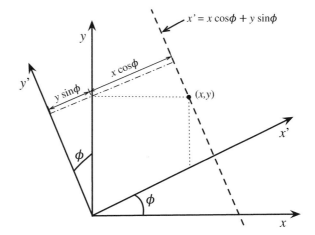

Fig. 8.3 Objection being imaged using CT, for use in Exercise 8A

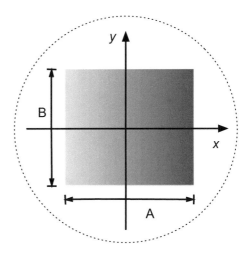

an attenuation profile of $\mu(x, y) = ax + b$. Derive an expression for the projection when $\varphi = 0°$ and $\varphi = 90°$.

8.2   The Sinogram

The different projections defined by Eq. 8.1 may be represented visually as a sinogram, Fig. 8.4. The sinogram plots the intensity distribution, which we have called λ, along x' on one axis, with all possible orientations represented along the other axis. The name arises from the sinusoidal shapes mapped out in this representation that arises from the circular nature of the acquisition process, something you will explore in Exercise 8B.

Fig. 8.4 Example of a sinogram that arises from plotting the projected intensities along the projection plane against projection angle

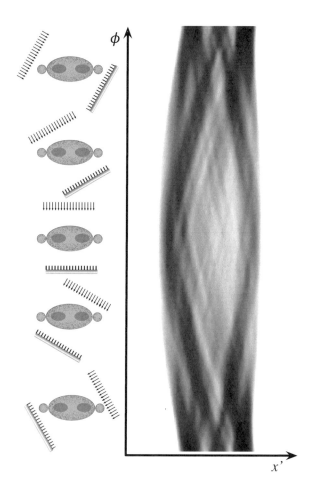

Fig. 8.5 Geometry for a
point source being images
using SPECT, for use in
Exercise 8B

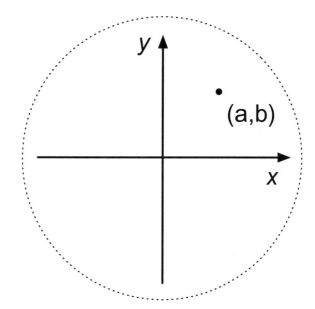

Exercise 8B A SPECT system is being used to image a point source at $x = a$ and $y = b$, as shown in Fig. 8.5.

(a) Show that the locus mapped out on a plane defined by x' (the projection axis) and ϕ (the projection angle) is given by:

$$x' = \sqrt{a^2 + b^2} \sin\left(\phi + \tan^{-1}\frac{a}{b}\right)$$

(b) Plot a number of loci for different a and b on a graph of x' versus ϕ. Hence, explain what this plot might look like when a general object is imaged and why it is called a sinogram.

8.3   Slice-Projection Theorem

The projection formulation is formally an example of the Radon Transform and we can relate this directly to the more familiar Fourier transform. Consider a projection $\varphi = 0$:

$$\lambda_0(x') = \iint\limits_{-\infty}^{\infty} \mu(x, y)\delta(x - x')\mathrm{d}x\mathrm{d}y = \int\limits_{-\infty}^{\infty} \mu(x', y)\mathrm{d}y \qquad (8.4)$$

If we do a 1D Fourier transform:

$$\Lambda_0(u) = \int \lambda_0(x)^{-2\pi jux}\mathrm{d}x = \iint \mu(x, y)\,\mathrm{e}^{-2\pi jux}\mathrm{d}x\mathrm{d}y \qquad (8.5)$$

This looks like a 2D Fourier transform if we rewrite it as:

$$\Lambda_0(u) = \iint \mu(x, y)\mathrm{e}^{-2\pi j(ux+vy)}\mathrm{d}x\mathrm{d}y\Big|_{v=0} = \mathrm{M}[u, 0] \qquad (8.6)$$

i.e. this equation is a special case of the 2D Fourier transform when $v = 0$. This suggests that the 1D Fourier transform of λ_0 gives the 2D Fourier transform of μ along a particular line. We might say more generally then:

$$\Lambda_\phi(u) \Leftrightarrow \mathrm{M}[u, v] \qquad (8.7)$$

i.e. there is a direct relationship between the 1D Fourier transform of the projection and the full 2D Fourier transform of the image. This result is called the slice-projection or central-slice theorem (or in fact any number of variations on these or similar words). What it implies is that if we gather a series of projections (and Fourier transform them), then we can reconstruct the 2D Fourier transform of our object and thus to reconstruct our image, all we would need to do is an inverse (2D) Fourier transform. It follows that we would need a full 180° rotation and we would need 'enough' projections to populate the Fourier domain sufficiently to allow for a good reconstruction. We have considered here only parallel projections, systems often operate with a fan beam which makes the maths a little more complex, but the principles are the same.

8.4 Filtered Back-Projection

The derivation above implies that we need to do an inverse Fourier transform, this is not necessarily very efficient in practice, and there is a more practical solution, based on the same principles. Starting with the idea that we need to do an inverse Fourier transform to reconstruct $\mu(x, y)$:

$$\mu(x, y) = \mu(r, \theta) = \int\limits_0^\pi \int\limits_{-\infty}^{\infty} \mathrm{M}(\rho, \phi)\mathrm{e}^{2\pi j\rho(x \cos\phi + y \sin\phi)}|\rho|\mathrm{d}\rho\mathrm{d}\phi \qquad (8.8)$$

where we are making use of the inverse Fourier transform in polar frequency coordinates (see Chap. 6) but writing the spatial coordinates in terms of the equivalent Cartesian coordinates.

If we now define $x' = x \cos\theta + y \sin\theta$ (as we did for Eq. 8.3), we can separate the two integrals and write this as:

$$\mu(x, y) = \int_0^\pi \lambda_\phi^\dagger(x') \mathrm{d}\phi \tag{8.9}$$

where we have defined a new quantity:

$$\lambda_\phi^\dagger(x') = \int_{-\infty}^\infty M(\rho, \phi)|\rho|e^{2\pi j\rho x'} \mathrm{d}\rho \tag{8.10}$$

This expression is of the form of a 1D (inverse) Fourier transform of the product of M and $|\rho|$ with respect to ρ. This is the same as the convolution of the (inverse) Fourier transform of the two separately, i.e.:

$$\lambda_\phi^\dagger(x') = i\mathrm{FT}(M)i\mathrm{FT}(|\rho|) \tag{8.11}$$

At this point, we can look back to Eq. 8.7 and write:

$$\lambda_\phi^\dagger(x') = i\mathrm{FT}(\Lambda_\phi)i\mathrm{FT}(|\rho|) \tag{8.12}$$

We know what the inverse Fourier transform of Λ_φ is and so we get:

$$\lambda_\phi^\dagger(x') = \lambda_\phi(x')i\mathrm{FT}(|\rho|) \tag{8.13}$$

This finally tells us that this new thing $\lambda_\phi^\dagger(x')$ is a 'filtered' version of the projection data that we have acquired, with $|\rho|$ defining the frequency response of the filter. The overall filtered back-projection process is thus to take the projection data, filter it by applying the equation above and the sum over all the projection angles according to Eq. 8.9. In practice, it is more efficient to swap the order and do the back-projection first, followed by filtering.

The only problem with this relatively simple process is the filter function itself. You will notice that its magnitude tends to infinity as ρ tends to infinity; this prevents us from computing the inverse FT. In fact, this function implies that back-projection is an ill-posed problem: the higher the frequency, the more the filter magnifies the amplitude. Since the filter amplifies high frequencies, we would expect it to amplify the noise in the reconstructed images.

In practice, the data will be band limited by the detection system and thus, we might instead use:

$$P(\rho) = \begin{cases} 0 & |\rho| \geq \rho_{max} \\ |\rho| & |\rho| < \rho_{max} \end{cases} \tag{8.14}$$

For which an inverse exists. This still might have undesirable effects when applied to real noisy data. Other filters exist, including the popular Shepp-Logan filter that has better rejection of high frequencies:

$$P(\rho) = \left| \left(\frac{2\rho_{max}}{\pi} \right) \sin \left(\frac{\pi \rho}{2\rho_{max}} \right) \right| \tag{8.15}$$

8.5 SPECT

For SPECT, we do something similar to CT and rotate an array of detectors around the subject on a gantry. The detector has a collimator on the front to ensure we only detect gamma rays arriving from a precise direction. We can write the projection as:

$$\lambda_\phi \left(x' \right) = \int_{AB} f(x, y) \mathrm{d}y' \tag{8.16}$$

where the received intensity is now the integral of all the emissions, f, along a line AB through the object. As with the CT reconstruction problem, we can apply back-projection methods to the projection data we acquire. Note that this equation does not consider the attenuation of the gamma rays, this is something we have to correct for, but it is not a simple as attenuation of X-rays, as the gamma rays all originate at different locations within the body and thus will experience different attenuation dependent upon their distance from the detector. The main limiting factors on resolution in SPECT are the size of the detectors and scatter that cannot be corrected for by the application of collimators.

8.6 PET

The added information we get with PET is that each emission gives rise to two gamma photons launched at precisely 180° to each other. Thus, by pairing up detections, we can more precisely locate the line on which the emission occurred—the line of response (LoR). To do this, we need a ring of detectors mounted around the subject to simultaneously record all emissions that arise within the body.

PET reconstruction relies upon coincidence detection: to determine the line of response, we need to identify the two detectors at which two separate photons from the decay process have arrived. PET systems typically operate with time windows of the order of 12 ns. Any two photons received within that window are presumed to have

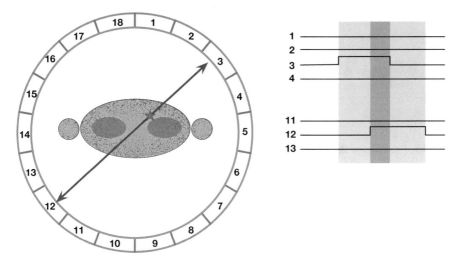

Fig. 8.6 The principle of coincidence detection. An event is recorded as having occurred on the LoR connecting detectors 3 and 12 based on signals being generated by both within the specified time window

come from the same emission event—are coincident—and thus an emission must have occurred along the LoR that connects those two detectors. This is illustrated in Fig. 8.6, where an event is recorded as having occurred on the LoR between sensors 3 and 12 based on an overlap in the signals generated in both detectors in response to a received photon (which has been converted to a signal of duration τ seconds).

The series of events recorded by the PET scanner is often called listmode data, from which the sinogram can be constructed. Again, back-projection methods can be used to reconstruct the image. Since PET uses coincidence detection, we are able to locate emissions to a given line of response with much more precision than in SPECT projections, leading to higher resolution. Even better resolution can be achieved using time-of-flight information, where the timing differences between gamma photons arriving at the detectors are used to estimate the location of the emissions along the LoR. A number of processes still limit the spatial resolution in PET.

8.6.1 Randoms

Accidental, or random, coincidences are events where two disintegrations occur very closely together in time. This can lead to the wrong LOR being chosen as shown in Fig. 8.7 (left). Note that scattered photons can cause a similar problem as illustrated in Fig. 8.7 (right). Correction methods typically try to estimate the number of accidental coincidences that are occur and subtract this from the acquired data, but this cannot overcome the 'blurring' of the image that results from incorrectly assigned LoR.

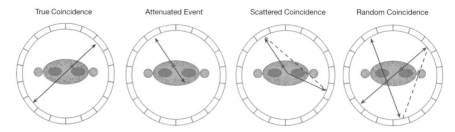

Fig. 8.7 Examples of the incorrect assignment of the LOR

8.6.2   *Positron Range*

An important limiting factor in PET systems is the need for the positron emitted by the radionuclide to annihilate with an electron. This will happen some distance away from the original emission and will vary from positron to positron, in part determined by properties of the tissue. This has a blurring effect on the final reconstructed image increasing the width of the PSF.

Further Reading
For more information on CT imaging see

Introduction to Medical Imaging, Nadine Barrie Smith and Andrew Webb, Cambridge University Press, 2011, Chapter 2.
Medical Imaging: Signals and Systems, 2nd Ed, Jerry L Prince and Johnathan M Links, Pearson, 2015, Chapter 6.
Webb's Physics of Medical Imaging, 2nd Ed, M A Flower (Ed), CRC Press, 2012, Chapter 3.

For more information on PET and SPECT imaging
Introduction to Medical Imaging, Nadine Barrie Smith and Andrew Webb, Cambridge University Press, 2011, Chapter 3.
Medical Imaging: Signals and Systems, 2nd Ed, Jerry L Prince and Johnathan M Links, Pearson, 2015, Chapter 9.
Webb's Physics of Medical Imaging, 2nd Ed, M A Flower (Ed), CRC Press, 2012, Chapter 5.

Chapter 9
Fourier Reconstruction—MRI

Abstract In this chapter, we will consider Fourier-based reconstruction as is found in magnetic resonance imaging (MRI). We will see how gradient fields are used to encode the signals with spatial information and how a Fourier transform can then be used to recover a 2D or 3D image. We will explore some of the surprising consequences of this for image resolution and field of view.

An MRI scanner consists of the following principal components:

- Main magnet
- Set of switchable gradient coils
- RF coils (and power amplifiers)
- Pulse sequence and receive electronics

When looking at NMR in Chap. 5, we saw why we need the magnet, the RF coils and pulses. However, we have yet to explore what the gradient coils do, and they are key to spatially resolving the signals so that we can generate images.

9.1 Gradients

In Chap. 5, we considered a B_0 field that points along the z-axis and it has the same in magnitude everywhere. The gradient coils can alter the magnitude of B_0 and this is the key to spatial encoding. With this, we can spatially select slices, or even pixels within a slice, paving the way to imaging.

Generally, there are three orthogonal gradient coils (in x, y, z). Each gradient coil can be used to add or subtract a spatially dependent magnetic field to the main field:

$$\boldsymbol{B} = \left(B_0 + G_x x + G_y y + G_z z\right)\hat{z} = (B_0 + \boldsymbol{G} \cdot \boldsymbol{r})\hat{z} \tag{9.1}$$

Importantly, the direction of the field along z, the main scanner axis, is maintained. Only the *magnitude* is changed. The gradient applied, $\boldsymbol{G} = \left(G_x, G_y, G_z\right)$, is measured in mT m^{-1} (or Gauss cm^{-1}) and maximum values around 40 mT m^{-1} are typical for clinical systems. As we will see, these coils must be switched on/off very

© Springer Nature Switzerland AG 2019
M. Chappell, *Principles of Medical Imaging for Engineers*,
https://doi.org/10.1007/978-3-030-30511-6_9

rapidly (0.1–1 ms). The so-called slew rate indicates the maximum rate of change of gradient values, of the order of 5–250 mT m^{-1} s^{-1}. Magnetic field switching is limited to limit induction of eddy currents in the patient.

The principle of the NMR experiment is to excite spins at their Larmor (resonant) frequency using RF waves. The gradient coils allow us to encode the spatial position of the NMR signals as part of this process. The trick is to use the Larmor frequency and the phase of the transverse magnetisation to form an actual MR image.

9.2 Slice Selection

The first thing we do is select a specific slice, so that the received signal only arises from that selected region of tissue. If we apply a gradient $G = (G_x, G_y, G_z)$, this results in a Larmor frequency of:

$$\omega(r) = \gamma (B_0 + G \cdot r) \qquad (9.2)$$

If we want to restrict ourselves to a single slice located at z, we would apply a gradient of $G = (0, 0, G_z)$ to obtain a Larmor frequency dependent on z:

$$\omega(z) = \gamma (B_0 + G_z z) \qquad (9.3)$$

As long as we only excite and receive signals at the precise frequency corresponding to a specific value of z, then in principle, we could select an infinitesimally thin slice of the object. In reality, this is neither possible nor desirable, since we can neither deliver such a precisely defined RF excitation (a real RF pulse has a finite bandwidth) nor get enough signal if we only excite an infinitely small section of tissue since the signal magnitudes depend upon the number of protons involved.

Rather than applying a pure sinusoidal waveform at one particular frequency, a waveform that excites a range of frequencies is applied, which in turn excites a range of tissues, i.e. a thick slice or 'slab', as shown in Fig. 9.1. By applying RF over the frequency range [ν_1, ν_2], a smaller z-gradient, G_1, or a larger z-gradient, G_2, will cause excitation in a thicker or thinner slice, respectively. The combination of gradient and frequency being used to select the position of the slice.

In practice, we use three parameters for slice selection:

- z-gradient strength G_z,
- RF centre frequency $\bar{\omega} = (\omega_1 + \omega_2)/2$, and
- RF frequency range (bandwidth) $\Delta\omega = |\omega_2 - \omega_1|$.

With these parameters, it is possible to control slice position and slice thickness, as you will show in Exercise 9A.

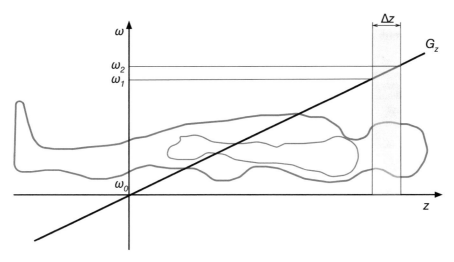

Fig. 9.1 Example of the use of a gradient (in z) and selective excitation to achieve slice selection in MRI

Exercise 9A Derive the following relationships between slice thickness and (mean) slice position:

$$\bar{z} = \frac{\bar{\omega} - \omega_0}{\gamma G_z}, \quad \Delta z = \frac{\Delta \omega}{\gamma G_z}$$

Calculate the required RF bandwidth to select a 5 mm slice in a clinical MRI scanner at 3 T with maximum gradient field strength of 40 mT m^{-1}.

If we want to image a volume, we can simply acquire multiple slices sequentially, but with the cost of extending the total acquisition time in proportion to the number of slices we want. The thickness of the slice we choose will, in practice, be mainly limited by how much tissue (i.e. water in tissue) we need to get sufficient SNR, and crucially CNR, for the application we are interested in. A thicker slab resulting in higher SNR at the cost of poorer resolution in the 'slice direction'. One consequence of slice selection is that the slice thickness and thus resolution in the slice direction is independent of that within the slice (which is set by the frequency and phase encoding). It is not unusual to find MRI data with a different slice thickness to the dimensions of the voxel within the slice as a means of getting better SNR without losing too much apparent resolution within the slice.

A consequence of slice selection being reliant on an RF pulse containing a specific range of frequencies is that, in practice, the edges of the slice will not be perfectly defined. Since, we cannot, in a single pulse of RF, deliver only the frequencies we need to define the slice. There will also be some extra components associated with the short duration of the pulse, and thus the pulse shape. For this reason, it is common to find that a small gap may be left between slices when doing multi-slice MRI and that a certain amount of optimisation of RF pulse shape is required for high quality MRI.

Exercise 9B By considering the Fourier transform of a rectangular function, explain why it would be impossible to generate a slice selection RF pulse that performs selection in a slice with perfectly defined edges.

9.3 Frequency Encoding

The slice selection gradientallows us to select a slice, but we still need to be able to distinguish different locations within the slice if we are going to construct an image. Recall from Chap. 5 that the transverse magnetization of a sample after a pulse is given by:

$$M_{xy}(t) = M_{xy}(0^+) e^{-j(2\pi\omega_0 t - \phi)} e^{-t/T_2} \tag{9.4}$$

where

$$M_{xy}(0^+) = M_z(0^-) \sin\alpha \tag{9.5}$$

is the transverse magnetization immediately after excitation, i.e. application of a flip angle of α. The received signal is an integral over the slice, i.e. all of the tissue that was excited:

$$s(t) = \iint_{-\infty}^{\infty} A M_{xy}(x, y, 0^+) e^{-j2\pi\omega_0 t} e^{-t/T_2(x,y)} dxdy \tag{9.6}$$

where A is a scanner-specific gain (related to the receive electronics etc.), we are assuming ϕ is zero for simplicity and we are explicitly noting that M_{xy} and T_2 are varying spatially depending upon the tissue present within the slice.

If we lump together all the terms that relate to the object (and include the gain), we can write the received signal as:

$$s(t) = e^{-j2\pi\omega_0 t} \iint_{-\infty}^{\infty} f(x, y) dxdy \tag{9.7}$$

With the 'effective spin density' given by:

$$f(x, y) = A M_{xy}(x, y, 0^+) e^{-t/T_2(x,y)} \tag{9.8}$$

Thus, the demodulated signal (baseband) that we will get is:

$$s_0(t) = e^{+j2\pi\omega_0 t}s(t) = \iint\limits_{-\infty}^{\infty} f(x, y)\mathrm{d}x\mathrm{d}y \qquad (9.9)$$

This is a constant, all spatial dependency has been integrated out, and we have lost all the position information. To achieve spatial encoding, we need to apply a gradient during the FID.

What we need is a readout gradient that is applied orthogonally to the slice selection gradient. We will use x for simplicity:

$$\omega(x) = \gamma(B_0 + G_z x) = \omega_0 + \gamma G_x x \qquad (9.10)$$

We still receive a signal that is the integral over all the spins within the selected slice, but the spatially dependent Larmor frequency makes all the difference:

$$s(t) = \iint\limits_{-\infty}^{\infty} AM_{xy}(x, y, 0^+)e^{-j2\pi\omega(x,y)t}e^{-t/T_2(x,y)}\mathrm{d}x\mathrm{d}y \qquad (9.11)$$

$$s(t) = \iint\limits_{-\infty}^{\infty} AM_{xy}(x, y, 0^+)e^{-j2\pi(\omega_0+\gamma G_x x)t}e^{-t/T_2(x,y)}\mathrm{d}x\mathrm{d}y \qquad (9.12)$$

$$s(t) = e^{-j2\pi\omega_0 t}\iint\limits_{-\infty}^{\infty} AM_{xy}(x, y, 0^+)e^{-j2\pi\gamma G_x tx}e^{-t/T_2(x,y)}\mathrm{d}x\mathrm{d}y \qquad (9.13)$$

Using again the definition of the effective spin density, $f(x, y)$, and demodulating the signal, we now obtain:

$$s_0(t) = \iint\limits_{-\infty}^{\infty} f(x, y)e^{-j2\pi\gamma G_x tx}\mathrm{d}x\mathrm{d}y \qquad (9.14)$$

This baseband signal, obtained by applying a readout gradient, looks like a 2D Fourier transform $F(u, v)$ of $f(x, y)$, with:

- Spatial frequency variable in the x-direction: $u = \gamma G_x t$
- Spatial frequency variable in the y-direction: $v = 0$.

Thus, as we take samples in t, we are in fact sampling the spatial frequency u. In MRI, we denote Fourier space as k-space and let $k_x = u$ and $k_y = v$, with k denoting the wave number.

9.4   Scanning k-Space

MRI essentially performs a 'scanning' of 2D Fourier space. The simplest case of frequency-encoded readout in u is shown in Fig. 9.2; the pulse sequence diagram on the left shows the actions being carried out by the scanner hardware. In this case, we firstly perform excitation which involves a RF pulse combined with slice selection gradient, and then we apply an x gradient during the readout period indicated by the ADC being set high. The longer we apply G_x the more u increases and thus, we take a path through k-space along the u-axis.

It is more useful if a readout gradient is first applied in $-G_x$ and then in $+G_x$, giving a frequency readout from $-u$ to u as shown in Fig. 9.3.

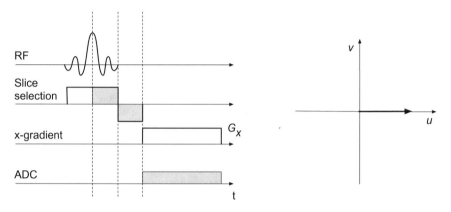

Fig. 9.2 Using a G_x readout gradient to acquire samples along u

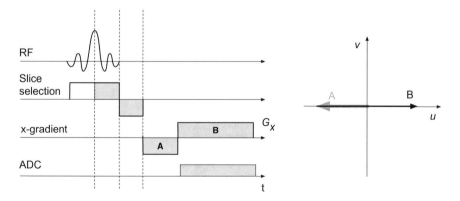

Fig. 9.3 Reading out the whole of the u-axis

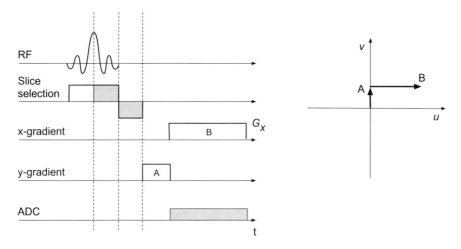

Fig. 9.4 Using a phase encoding gradient

So far, we have used frequency encoding as a mechanism to readout data in the u-direction. Phase encoding allows[1] us to first position the readout line in the v-direction:

$$s_0(t) = \iint\limits_{-\infty}^{\infty} f(x, y) e^{-j2\pi\gamma G_x x t} e^{-j2\pi\gamma G_y T_p y} \mathrm{d}x \mathrm{d}y \qquad (9.15)$$

Notice that the new term that depends upon y and the gradient G_y is not 'on' during the readout itself but occurs prior to it for a fixed period of time T_p, as shown in Fig. 9.4. We have used the G_y gradient to move us 'up' the v-axis first. The phase accumulated during the y-gradient pulse is given by $\phi_y = -\gamma G_y T_p y$.

The combination and phase and frequency encoding allow us to capture signals that describe from the whole of the 2D k-space. For example, we might follow the trajectory shown in Fig. 9.5.

Exercise 9C Sketch the pulse sequence diagram and k-space trajectory for the following readout gradient $G_x = G_y = G$, i.e.:

$$\omega(x, y) = \gamma(B_0 + Gx + Gy)$$

---

[1] The description in this book separates the two processes of frequency encoding and phase encoding, arguably both involve the phase of the received signal and (some would argue) should both be called phase encoding. But, it is somewhat easier to name them differently to emphasis their different roles in acquiring samples from k-space, and thus that convention has been adopted here.

Fig. 9.5 A k-space
trajectory to sample the
whole of 2D k-space from
one excitation

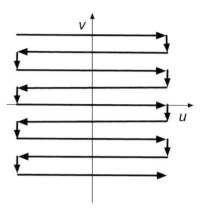

9.5   MRI Reconstruction

Once we have scanned k-space, the spatial signal can be recovered by applying an inverse Fourier transform to $F(u, v)$. For a discretely sampled k-space, this can be done very efficiently using the fast Fourier transform. Because the MRI instrument acquires information in the frequency domain and we (inverse) Fourier transform to get the image, there are some important implications for how we choose our sampling of k-space.

9.5.1   Field of View

This is where we need to think carefully about sampling theory. As we saw in Chap. 6, when sampling a signal we are in effect performing a multiplication of the signal with a train of delta functions, these having some regular spacing associated with the sampling rate. This process gives rise to replicated copies of the frequency spectrum of the signal in the frequency domain. The closer the samples are together, i.e. a higher sampling rate, the further apart these replicas appear in the frequency spectrum of the sampled signal.

These same principles hold for the case of MRI reconstruction only now we are *sampling in the 2D frequency domain*, i.e. it as if we were taking the true 2D (or 3D) frequency spectrum of the object and multiplying by a grid of delta functions. This has the effect of creating replicas of the true image in the image domain; the closer the samples are together the further apart these replicas become. This means that we have to ensure we are sampling k-space densely enough otherwise we get wrap-around artefacts as shown in Fig. 9.6.

Formally, The Nyquist theorem tells us that we have to sample at a rate twice the highest frequency in the signal, which will be related to the bandwidth of the signal, to avoid aliasing. For MRI, we need to define the 'bandwidth' in the reconstructed

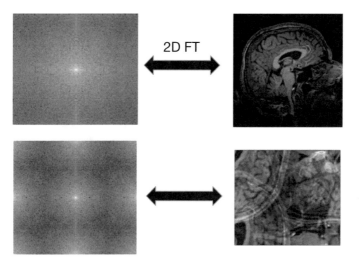

Fig. 9.6 The effect of under-sampling (aliasing) in MRI. The top row is 'fully' sampled, in the bottom row every other line in k-space has been removed (factor of 2 under-sampling), resulting in overlapping in the reconstructed image

image, which is the FoV. Just as the sampling rate we apply to a signal defines the separation between replicas in its frequency spectrum, the k-space sampling interval in MRI defines the FoV in the spatial domain: $\text{FoV}_x = \frac{1}{\Delta k_x}$ and $\text{FoV}_y = \frac{1}{\Delta k_y}$.

To avoid wrap-around artefacts, we have to ensure that the FoV surrounds all of the physical objects we are imaging or that we use gradients with the excitation to only select a specific volume of tissue. Any tissue outside the box defined by our k-space sampling interval will end up appearing inside the imaging volume, overlapping with the tissue that is meant to be there. For MRI, this is the equivalent to applying an anti-aliasing filter to a signal to remove high frequencies that would otherwise get aliased into the sampled signal.

9.5.2 k-*Space Coverage and Resolution*

We also have a choice as to how much of k-space we will acquire. Most critically, we have to decide what are the highest frequencies we are going to be interested in. The higher the frequencies we want, the more time we will have to spend in the readout part of the sequence. Figure 9.7 shows examples of what happens when we restrict ourselves to only part of k-space, the implications of this are similar to what we already know from Fourier transforms in 1D. Namely, if you throw away higher frequencies, you lose fine details in your image. You might appreciate that the most important part of k-space that we must sample is the centre, as most of the

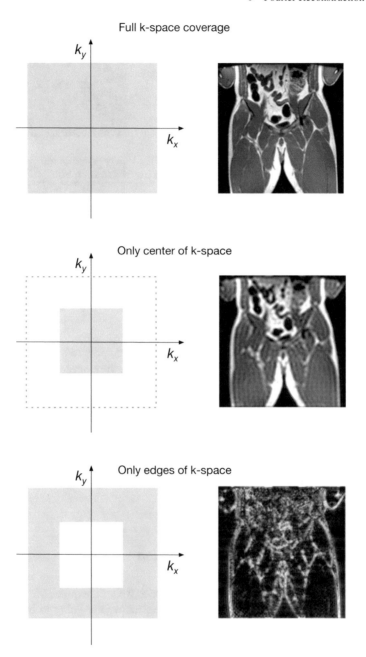

Fig. 9.7 Effect of the choice of k-space coverage on reconstructed MR image

information in the image is represented there. The effective resolution of the image will be directly related to the highest frequency acquired: $\Delta x = 1/(2k_{max})$.

9.5.3 The Effect of T_2 Decay

As we have seen acquiring samples of k-space takes some time, since it is samples from the signal post-excitation that are used to populate k-space. During this time, both T_1 and T_2 decay processes are occurring. Since we are sampling the transverse magnetization, it is the T2 decay that matters, and this will be reducing the amplitude of the (demodulated) signal during the readout. In Exercise 9D, you are asked to explore the effect this will have on the reconstructed image. In practice, if we want to avoid these effects and achieve a more faithful reproduction of the true object, we need to keep the acquisition as short as possible. One solution is to segment the acquisition. Although it is possible to acquire the whole of k-space from a single excitation by using a combination of frequency and phase encoding to 'steer' through k-space; in practice, it may be preferable to use multiple excitations and from each one, acquire only a part, e.g. a single line, of k-space. The cost being that the whole acquisition will take longer.

> **Exercise 9D** If the readout was to take a 'long' time what would the effect of transverse relaxation be on later k-space samples compared to early ones? What effect would this have on the final images if (as is conventional) the largest positive k_y lines are acquired first, with the largest negative last as in Fig. 9.5?

9.6 More Advanced k-Space Trajectories

9.6.1 Partial Fourier

The total time to acquire an image is determined by the time it takes to sufficiently sample k-space. We have already noted that to avoid effects from transverse decay during readout, we need to limit the time spend sampling after an excitation. A simple strategy that can help reduce the number of k-space samples required is simply to not collect the whole of k-space. This exploits the fact that for a real-valued image, there will be symmetries in k-space that make some of the data redundant. In theory, therefore, it is only necessary to collect half of k-space, e.g. only the lines of positive k_y. In practice, this only works when there are no other sources of error in the readout

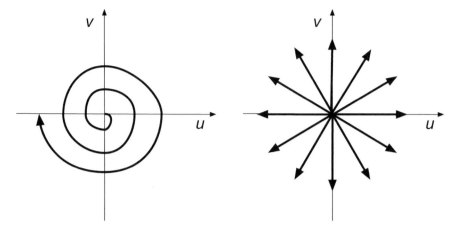

Fig. 9.8 Alternative *k*-space trajectories. The spiral (left) can be completed after a single excitation. The spokes (right) might be done from separate excitations; otherwise, it would be necessary to traverse a trajectory from the end of one spoke to the start of the next

and a common solution is to collect a slightly larger fraction including more lines around the centre of *k*-space, to allow for correction of errors.

9.6.2   Non-Cartesian

We are not restricted to only using a 'Cartesian' trajectory in *k*-space. We could instead use schemes like those shown in Fig. 9.8 including spiral or spokes. Such trajectories can offer specific advantages such as lower sensitivity to motion of the tissue during acquisition. They do bring the challenge that efficient Fourier transformation algorithms require data on an even grid and thus data from such trajectories is often resampled onto a grid before reconstruction.

9.6.3   3D k-Space

We noted in Sect. 9.5, one way to acquire a full 3D volume using MRI is to work slice by slice. We can excite one slice at a time and then do some form of 2D readout within the slice, before moving on to the next slice. A disadvantage of this scheme is that we only excite a small subset of the tissue (and thus the hydrogen nuclei in the water in the tissue) at a time, which limits the SNR. We can instead excite a much larger volume of tissue and use frequency encoding combined with phase encoding in two (e.g. *y* and *z*) directions to acquire a 3D *k*-space all in one excitation. This has the benefit of higher SNR since more hydrogen nuclei contribute to the measured signal.

The major disadvantage of such a scheme is that it takes longer to traverse the full 3D *k*-space than it does to only acquire the *k*-space for a single 2D slice. The resulting reduction in true resolution of the image means that the PSF of 3D is normally poorer than multi-slice 2D due to the T_2 decay effects we met in Sect. 9.5.

9.6.4 Compressed Sensing

All of the sampling approaches we have considered up to this point have assumed that we completely sample the whole of k-spare or can efficiently 'fill-in' any bits we miss. This is based on the demands of the Nyquist sampling theorem. If we did not have sufficient samples this would lead to artefacts in the reconstructed image. In reality, most images contain a lot of redundant information; this is why image (and video) compression techniques are so effective. This implies that in theory, we might not need to collect all of the *k*-space data to still have all of the information needed to reconstruct the image. But to do so, we would no longer be able to simply use a Fourier-based reconstruction algorithm.

Compressed sensing is an approach that follows these principles. In the ideal case, a compressed sensing acquisition involves random sampling for k-space. This process would give rise to artefacts in the reconstructed images, but because the samples are random, these artefacts are not coherent, i.e. they would appear somewhat like noise. The compressed sensing reconstruction uses some form of added constraint in the reconstruction to extract the image from amongst this noise that exploits known properties of the image. The most widely used constraint being the sparsity of the image under some transformation. For example, an angiographic image of blood vessels is a visibly sparse image; in the extreme case, voxels are either background (no intensity) or vessel. The image can be represented very compactly in terms of finite differences, and this can be used as a 'sparsifying transform' to constrain the reconstruction. Compressed sensing reconstructions are performed as part of a reconstruction algorithm, something we will consider more in Chap. 10. In practice, it is not possible to truly randomly sample *k*-space, because, as we have seen, frequency and phase encoding require us to take a 'path' through *k*-space. However, the principles of compressed sensing can still be applied to under-sampled MRI data.

Further Reading
For more information on Magnetic Resonance Imaging see

Introduction to Medical Imaging, Nadine Barrie Smith and Andrew Webb, Cambridge University Press, 2011, Chapter 5.
Introduction to Functional Magnetic Resonance Imaging, 2nd Ed, Richard B Buxton, Cambridge University Press, 2009, Chapter 4.

Medical Imaging: Signals and Systems, 2nd Ed, Jerry L Prince and Johnathan M Links, Pearson, 2015, Chapter 13.
Webb's Physics of Medical Imaging, 2nd Ed, M A Flower (Ed), CRC Press, 2012, Chapter 7.

Compressed sensing for MRI is introduced in
Lustig, M., Donoho, D. & Pauly, J. M. Sparse MRI: The application of compressed sensing for rapid MR imaging. *Magnetic Resonance in Medicine,* 58, 1182–1195 (2007).

Chapter 10
Principles of Reconstruction

Abstract In this chapter, we will think more generally about the principles of image reconstruction that we have met in various guises in Chaps. 7–9. Having first defined a general analytic form for reconstruction, we will go on to consider how iterative methods can be used and where these offer advantages for particular imaging techniques.

10.1 Analytic Reconstruction

In Chap. 6, we considered the general image formation problem, whereby we sought to describe the effect of the imperfections of our complete imaging system in the form of a convolution with a point spread function.

$$g(x, y) = h(x, y) \otimes f(x, y) \tag{10.1}$$

Or more generally

$$g = H\{f\} \tag{10.2}$$

When we are dealing with so-called natural images, e.g. photographs, we effectively measure $g(x, y)$ directly. We noted in Chap. 6 that if we were to attempt to get a more faithful representation of the object from the imperfect image, we would perform image restoration

$$\hat{f} = H^{-1}\{g\} \tag{10.3}$$

where H^{-1} is the inverse operation to H.

As we have seen in Chaps. 2–5, in medical imaging, we normally measure some other quantities related to the object of interest, such as projections or frequency information. In a general sense, we might thus say we actually have

$$\Lambda(\varsigma) = P\{\varsigma, f(x, y)\} \tag{10.4}$$

M. Chappell, *Principles of Medical Imaging for Engineers*,
https://doi.org/10.1007/978-3-030-30511-6_10

where $\Lambda(\varsigma)$ is the ideal measurement from the imaging device for a particular set of imaging parameters ς, and P is an operator that describes the acquisition process. For example, $\Lambda(\varsigma)$ might be projections from a CT system over projection angles, or measurements from an MRI system across 2D (or 3D) k-space. For the reconstruction methods we have considered in Chaps. 7–9, we have effectively sought to analytically invert the transform in Eq. 10.4, i.e.

$$g(x, y) = R\{\Lambda(\varsigma)\} \tag{10.5}$$

where we might call R the reconstruction operation. In the ideal case the operator P is invertible and thus $R = P^{-1}$, and we might hope to be able to reconstruct the original image (absent any contribution from noise). We might call this the general image reconstruction problem. This is separate from the idea of the general image restoration problem, but we can see that the two are linked as in Fig. 10.1.

In reality, what we measure using the imaging system will be an imperfect representation of the true data due to imperfections in the acquisition. Additionally, we might not be able to exactly perform the inverse of the transformation P, but only some approximation to it. Hence, post-reconstruction, rather than recovering $f(x, y)$ the 'true' image of the object, we get $g(x, y)$; and thus, it is the combination of the imperfections in the data collection *and* the reconstruction method that give rise to the non-ideal PSF.

For example, for frequency reconstruction, R is the Fourier transformation of the 2D (or 3D) k-space data. The process of applying the Fourier transform exploits the fast Fourier transform (FFT) algorithm since we are working with sampled k-space data, not analytical functions. This requires the data to be sampled evenly across a

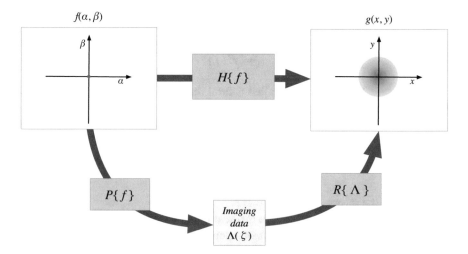

Fig. 10.1 Idea of image formation, encapsulated by the operator H (e.g. the PSF), is linked to that of image acquisition, the acquisition function P, and image reconstruction, the reconstruction function R

grid of points in k-space. Where this is the case, then the analytical reconstruction process can be performed exactly. However, as we saw in Chap. 9, there are situations where we do not sample the data in this way, for example, radial or spiral k-space trajectories. For such data, it is generally necessary to first resample the measured data on to a grid, before the FFT is applied. Thus, the reconstruction process becomes inexact. Additionally, we have seen that T_2 decay during the readout, that reduces the signal intensity, manifests as a blurring of the final image.

For tomographic reconstruction, P represents the process of taking projections, and in Chap. 8, we derived the inverse in the form of back-projection. When we subsequently defined filtered back projection, we discovered that the ideal 'filter' could not be realized in practice (and in fact was ill-posed). Instead, we chose a suitable filter function, and thus an approximation to R, that suppresses noise at the expense of increasing blurring in the reconstructed images.

10.2 Reconstruction as Cost Function Minimization

For all of our imaging methods, particular problems or challenges arise when we try to perform the reconstruction operation R. However, we can generally define a suitable form for P: the 'forward' relationship from object to measured imaging data. An alternative to the analytic reconstruction approaches we have considered thus far is to think of reconstruction as the process of minimizing the error, or more generally a cost function, between the measured data and a prediction of that data based on a guess of what the real image should look like

$$\text{Cost} = \text{cost}(\Lambda(\varsigma), P(\varsigma, g(x, y))) \tag{10.6}$$

We might, for example, take as the cost function (for images that take real values) the sum of squared differences over all the measurements.

$$\text{Cost}_{\text{SSD}} = \sum_{\text{measurements}} (\Lambda(\varsigma) - P(\varsigma, g(x, y)))^2 \tag{10.7}$$

Once we have defined a suitable cost function, we can then use an optimization algorithm to minimize the cost function and thus find a best prediction for the image $g(x, y)$. This would typically involve making some initial guess of the image as a starting point, for example, using an analytical reconstruction method.

For discrete data, we can write the forward relationship between the object and the associated predicted measurements under P as

$$\Lambda_i = \sum_{j=1}^{N} P_{ij} f_j(x, y) \tag{10.8}$$

where we have assumed that the imaging process can be represented by a system matrix $\mathbf{P}$, which describes the contribution to the ith measurement from the jth voxel,[1] i.e.

$$\Lambda = \mathbf{Pf} \tag{10.9}$$

As you find in Exercise 10A, if we seek a least-square error for image reconstruction, then

$$\mathbf{g} = \mathbf{R}\Lambda = \left(\mathbf{P}^{\mathrm{T}}\mathbf{P}\right)^{-1}\mathbf{P}^{\mathrm{T}}\Lambda \tag{10.10}$$

where $\mathbf{R}$ is the reconstruction matrix. Note that this result uses the matrix pseudo-inverse to account for the (common) situation where $\mathbf{P}$ is not a square matrix. This is a general algebraic reconstruction algorithm. But, one that is not very practical to use due to the need to invert what will typically be a relatively large system matrix. In practice, therefore, an iterative scheme, such as Expectation-Maximization, will be preferable.

Note that in reality, the measurements will be noisy, i.e.

$$\Lambda = \mathbf{Pf} + \mathbf{n} \tag{10.11}$$

We can subject this expression to the same analysis as in Exercise 6C to show that if $\mathbf{P}$ is ill-conditioned, as it would be for back-projection, for example, then reconstruction leads to noise amplification. As you will show in Exercise 10A, when $\mathbf{n}$ is the result of white noise then to achieve a maximum-likelihood solution for the reconstructed image, we would use the SSD cost function.

Exercise 10A

(a) For the discrete image acquisition as given by Eq. 10.9, write down the appropriate SSD cost function involving the estimated image $\mathbf{g}$.

(b) By differentiating the expression and setting to zero, show that the SSD cost function is minimized by the reconstruction formula in Eq. 10.10, where $\mathbf{P}$ is a non-square matrix.

(c) Write down an expression for the likelihood $P(\Lambda|\mathbf{g})$ for reconstruction of noisy data like that in Eq. 10.11, when the data is corrupted by white noise, i.e. $\mathbf{n} \sim N(0, \sigma\mathbf{I})$.

(d) By considering the (natural) log of the expression for the Likelihood from part (c), confirm that the maximum-likelihood solution would be the same as using a SSD cost function.

---

[1] Like in Chap. 6, we are treating our object as being divisible into discrete elements that correspond to the voxels in the resulting image.

10.3   Iterative PET Reconstruction

For PET, the measurement array $\mathbf{\Lambda}$ is composed of the events recorded along each LOR. In a similar manner to that in Sect. 10.2, we can represent the data acquisition process in matrix form as

$$\mathbf{\Lambda} = \mathbf{Pg} + \mathbf{r} + \mathbf{s} \tag{10.12}$$

where $\mathbf{P}$ is the system matrix and represents the relationship between the counts recorded on the ith LOR of emissions from the jth voxel in the object, $\mathbf{r}$ is an array of the number of random events and $\mathbf{s}$ is an array of the number of coincidences recorded on each LOR, respectively. The simplest version of the system matrix would be to use it to represent the Radon transform. However, there are a range of other processes that, as we saw in Chaps. 4 and 8, also affect the measured data. These can all be incorporated into the system matrix. This ability to model the various sources of error in the image acquisition process has made iterative reconstruction increasingly popular for PET. One approach is to assume that it is possible to separate the effects into different factors:

$$\mathbf{P} = \mathbf{P}_{\text{geom}} \mathbf{P}_{\text{attn}} \mathbf{P}_{\text{det}} \mathbf{P}_{\text{positron}} \tag{10.13}$$

where

- $\mathbf{P}_{\text{geom}}$ describes the mapping from image to scanner data. This might be calculated by taking line integrals through the voxel grid along each LOR to calculate the relative contributions of each voxel to the LOR, which can be done using ray-tracing algorithms found in computer graphics. This assumes that each detector can be represented as a point, whereas, in practice, a detector has a finite surface area. More sophisticated methods take the volume of intersection into account.
- $\mathbf{P}_{\text{attn}}$ models the attenuation of the gamma rays for a given LOR.
- $\mathbf{P}_{\text{det}}$ is used to account for factors arising from the detectors associated with each LOR such as dead time and blurring due to photon penetration and scatter.
- $\mathbf{P}_{\text{positron}}$ models the average blurring effect that arises from the distance the positron travels before annihilation. Compared to the other contributions, this is often small and can be ignored.

An advantage of this formulation is that although the system matrix itself is large (the number of LOR by the number of voxels), these individual matrices are likely to be sparse and thus highly compressible, allowing for more efficient computation.

We noted in Chap. 4 that positron emissions, being a form of radioactive decay, are a random process. This can be modelled by a Poisson distribution

$$\Lambda_i \sim \text{Poisson}(\langle \Lambda_i \rangle) \tag{10.14}$$

i.e. the actual measured number of counts in the ith LOR arises from a Poisson distribution with expected (mean) value $\langle \Lambda_i \rangle$. Strictly, Eq. 10.12 gives the expected counts detected along each LOR.

$$\langle \Lambda \rangle = \mathbf{Pg} + \mathbf{r} + \mathbf{s} \tag{10.15}$$

We can then interpret the elements of the system matrix as the probability that a pair of photons originating in a given voxel is detected along a specific LOR. Note that this idea of modelling the randomness inherent in PET imaging is similar to modelling the noise in Sect. 10.1, the difference being that when we modelled noise previously, we were assuming additive white noise.

As you will show in Excercise 10B, we can write the cost function, based on the log-likelihood, for PET reconstruction as

$$L = \sum_{i=1}^{N_{\mathrm{LOR}}} \left[\Lambda_i \ln \left(\sum_{j=1}^{N_v} (P_{ij} g_j) + r_i + s_i \right) - \sum_{j=1}^{N_v} P_{ij} g_j + r_i + s_i \right] \tag{10.16}$$

where the number of LOR is N_{LOR} and the number of voxels we are trying to reconstruct is N_v. Typically, this is maximized using some form of Expectation-Maximization iterative procedure; hence, iterative PET reconstruction algorithms are often called maximum likelihood expectation maximization (MLEM). Note that the unknowns include not only the desired image data $\mathbf{g}$, i.e. the estimated image $g(x, y)$, but also the number of random and scattered events detected along each LOR, $\mathbf{r}$ and $\mathbf{s}$, respectively. Often estimates for these are calculated prior to running the MLEM reconstruction. Estimation of scatter events ideally requires an estimate of the activity, i.e. $g(x, y)$. Hence, updating of this term during the iterations might be included as part of the reconstruction.

Exercise 10B

(a) Given that the counts on the ith PET LOR can be described by a Poisson process as in Eq. 10.14, write down an expression for $P(\Lambda_i | \langle \Lambda_i \rangle)$.

(b) Hence, write down an expression for the likelihood $P(\Lambda | \langle \Lambda \rangle)$, assuming that the counts detected along each LOR can be treated as independent and the number of LOR is N_{LOR}.

(c) Write down the expression for the log-likelihood. Identify which terms do not depend upon the estimated image $g(x, y)$ (nor the quantity of random or scatter events). Why can these terms be neglected when using this as a cost function for reconstruction.

(d) Using Eq. 10.15, find the expression for the cost function in Eq. 10.16.

10.4 Compressed Sensing MRI Reconstruction

Given the relative ease with which MR images can be reconstructed from sampled k-space data, the use of iterative reconstruction has not yet become so widely used compared to PET. Treating the reconstruction problem as one of maximizing a cost function does have a potential advantage where we have only a limited amount of k-space samples: under-sampled data. In Chap. 9, we noted that if we had fewer samples we either lost spatial detail by not sampling higher spatial frequencies, or we had more widely spaced samples which risked wrap-around artefacts in the reconstructed image. The latter effect was a result of the Nyquist sampling theorem that sets criteria on the density of samples needed in k-space for a given FOV.

Traditionally, MRI is constrained by the Nyquist theorem. This then determines the total time needed to acquire the k-space data with corresponding effect on scan time. To reduce scan time, we would need to reduce the number of samples acquired. One thing we might note about the final images we get from MRI is that they are compressible. For example, we could submit them to a compression algorithm, such as GZip, and without any loss of information obtain a version of an image that occupies far fewer bits. We could go further, e.g. JPEG, and with minimal loss of visible information compress the image to an even smaller number of bits. This implies there is some redundancy in the data, and at the same time some structure within the data we can use to simplify the representation. This can be exploited through the theory of compressed sensing.

The essential benefit of compressed sensing is that it allows us to sample below the limit set by the Nyquist theorem yet still recover the original data (or something close to it), as long as we can say something about the expected structure of that data. Compressed sensing for MRI requires that there is some (nonlinear) transformation that we can apply to the image that results in a sparse representation and that artefacts that arise from under-sampling are incoherent in the transform domain, i.e. the aliasing from under-sampling k-space appears random, noise-like, after the transform has been applied to the aliased image. The latter would ideally be achieved by randomly sampling points in k-space, but that is impractical and so many applications of compressed sensing use pseudo-random trajectories through k-space.

The application of compressed sensing requires the solution of an optimization problem of the form

$$\text{minimize} \quad \|\Psi\{g(x, y)\}\|_1$$
$$\text{subject to} \quad \|P\{\varsigma, g(x, y)\} - \Lambda(\varsigma)\|_2 < \epsilon \qquad (10.17)$$

The cost function we are attempting to minimize here is the 1-norm of the estimated image subject to some (sparsifying) transformation Ψ. For discrete images, the transform might take the form of a matrix that is multiplied by the vector of voxel values, in which case the 1-norm would be the sum over all the values in the resulting vector. This cost function minimization is subject to a constraint on the value of the 2-norm of the difference between the measured and predicted data. This follows the same rationale as the cost function minimization in Sect. 10.2 and ϵ controls the fidelity between the reconstruction and the measured data. Note that using a 2-norm for this term is the same as calculating the SSD for discrete data.

Various specialized methods exist to solve Eq. 10.17. A simple solution is to write the problem as

$$\text{minimize } \|P\{\varsigma, g(x, y)\} - \Lambda(\varsigma)\|_2 + \lambda\|\Psi\{g(x, y)\}\|_1 \qquad (10.18)$$

Which is an example of using a Lagrange multiplier. In essence, the parameter λ balances the two requirements of sparsity and fidelity, and needs to be chosen either manually or through some separate optimization processes.

A classic example of compressed sensing in MRI is of angiographic data. Here, the blood vessels are visualized, and these naturally appear sparse in the resulting image since only a fraction of the voxels contain a vessel and are thus nonzero. In this case, the sparse representation of the image can be the image itself, i.e. the sparsifying transform in matrix form $\Psi = \mathbf{I}$, the identity matrix. The optimization thus seeks an image that both matches the k-space data under a Fourier transform and also has as few nonzero voxels as possible. The 1-norm effectively counts the number of nonzero voxels in the image.

Further Reading

For more information on iterative reconstruction in CT imaging see

Webb's Physics of Medical Imaging, 2nd Ed, M A Flower (Ed), CRC Press, 2012, Chapter 3.

For more information on iterative reconstruction in SPECT/PET imaging see
Medical Imaging: Signals and Systems, 2nd Ed, Jerry L Prince and Johnathan M Links, Pearson, 2015, Chapter 9.
Webb's Physics of Medical Imaging, 2nd Ed, M A Flower (Ed), CRC Press, 2012, Chapter 5.

Compressed sensing for MRI is introduced in
Lustig, M., Donoho, D. & Pauly, J. M. Sparse MRI: The application of compressed sensing for rapid MR imaging. *Magnetic Resonance in Medicine, 58,* 1182–1195 (2007).

Part III
And Beyond ...

Chapter 11
Generating and Enhancing Contrast

Abstract In this chapter, we will consider further the principles of generating contrast in medical images. Having already explored physical principles that we can use to generate contrast between body tissues in earlier chapters, in this chapter, we will meet ways in which we can enhance the contrast and thus tailor our medical imaging devices to particular applications. We will consider ways to enhance the inherent contrast mechanisms available in different medical imaging techniques both through changes in the way the signals are acquired and through the use of externally introduced contrast agents.

We have now met a range of techniques that we can use to extract information from inside the body and have seen how we can reconstruct these signals into full 3D images. We noted in Chap. 1 that it is contrast between tissues and structures that are important and have seen that different methods are able to generate different degrees of contrast between different types of tissues. Very often, a particular imaging technique is chosen because it can produce a good contrast between particular tissues of interest. Thus, you can easily distinguish between bone and soft tissue using X-rays but are better using MRI to distinguish between different soft tissues. In some cases, the imaging methods do not provide the contrast we need on their own and we need to do something to boost the contrast. This can be achieved through the manipulation of the signal or the use of a contrast agent, or both.

> **Exercise 11A** Summarize the physical properties that the following imaging systems depend upon for the generation of contrast between tissues in the body (without the addition of a contrast agent):
>
> (a) CT
> (b) Ultrasound
> (c) MRI

© Springer Nature Switzerland AG 2019
M. Chappell, *Principles of Medical Imaging for Engineers*,
https://doi.org/10.1007/978-3-030-30511-6_11

11.1   Manipulating the Signal

All imaging systems depend upon particular physical processes to produce the signals which we then convert into images, and thus we are reliant upon differences in particular physical properties of the tissues to generate the contrast we need. For some of the methods we have met, there is not a lot we can do with the imaging equipment itself to change the contrast we can measure. We may do a lot to refine the sources and detectors to improve the SNR, increase the resolution or increase the speed of operation. But the contrast is predominantly set by the tissues we are imaging, we might choose a different frequency of operation to reduce and accentuate attenuation, but our influence will be limited at best. The one notable exception is MRI.

11.1.1   MRI Relaxation Contrast

In Chap. 5, we noted that the MR signal depends upon the proton density. Thus, we could generate a proton density contrast between tissues, or as it is commonly called a proton-density-weighted image. However, most tissues have a high and very similar proportion of water in them and thus the contrast in a proton-density-weighted image is not very large in many cases.

The other properties of the tissue we might exploit are the two magnetic relation constants T_1 and T_2. These both vary from tissue to tissue and have also been found to change in disease. To make an image that is sensitive to differences in T_2, we need to choose the time at which we measure the transverse magnetization M_{xy}. Depending upon how far along the T_2 decay process the signal is at the time of acquisition will mean that tissues with different T_2 values will have different intensities due to their differing decay rates. The time at which we measure M_{xy} is called the echo time (TE). The process of generating contrast between tissues by choosing the echo time is illustrated in Fig. 11.1. By choosing a nonzero echo time, we end up with an image whose intensities are related to the T_2 of the tissue but are not a direct measure of the T_2 constant: thus, they are termed T_2-weighted images.

Note that to achieve T_2 contrast, we have to wait for some signal decay to occur and that the choice of TE will be determined by the difference in T_2 we are hoping to observe. A consequence of this is that we do not have the maximum SNR possible compared to using a near-zero TE. We may well choose to use the spin echo that we met in Chap. 5 for T_2-weighted imaging since otherwise our images will strictly be $T_2{}^*$-weighted, we can recover the extra decay that gives rise to $T_2{}^*$ effects and thus get a purely T_2-weighted image that will have higher signal magnitude and thus better SNR. However, there are situations where a $T_2{}^*$-weighted image can be useful that we will meet in Chap. 12.

We can also generate a weighted image based on T_1, although we have to be a little more careful about how we design our NMR experiment. To achieve a T_1-based contrast, we vary how often we repeat the excitation: the repetition time (TR).

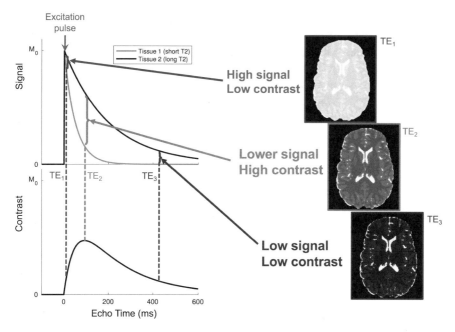

Fig. 11.1 Using echo time to generate contrast based on T_2 times. Reproduced with permission from Short Introduction to MRI Physics for Neuroimaging, www.neuroimagingprimers.org

If we repeat the excitation at a given TR, we eventually end up in steady state where the longitudinal magnetization of different tissues at the point of excitation will be different. Thus, when we excite, tipping the magnetization into the transverse plane where we can measure it, we will get a different intensity for each tissue. This is illustrated in Fig. 11.2 and is something you will explore in Exercise 11B.

Like T_2-weighting, this process of choosing the TR allows to generate an image whose intensities are related to the T_1 of the tissue in each voxel and is thus T_1-weighted, rather than being directly a measure of T_1.

Exercise 11B The longitudinal magnetization after application of a pulse with flip angle α, starting from an arbitrary initial magnetization M_{z0} is:

$$M_z = M_{z0} \cos \alpha + (M_0 - M_{z0} \cos \alpha)\left(1 - e^{-\frac{t}{T_1}}\right)$$

(a) Show that when $M_{z0} = M_0$ (the equilibrium magnetization of the tissue) this expression reduces to the T_1 recovery equation given in Chap. 5.

(b) Find an expression for the longitudinal magnetization immediately before the RF pulse (at the end of a TR) once the steady-state condition has been reached. Hence explain how we can achieve T_1-weighted contrast.

We have so far assumed that we will be using a flip angle of 90°, but in practice, this might not generate the largest signal. This is illustrated in Fig. 11.3, where a flip angle of 70° produces the largest longitudinal magnetization magnitude and thus the largest measurable signal. In general, the optimum, for a given T_1 and TR, is given by the Ernst angle, something you will derive in Exercise 11C:

$$\cos\theta = e^{-\frac{TR}{T_1}} \tag{11.1}$$

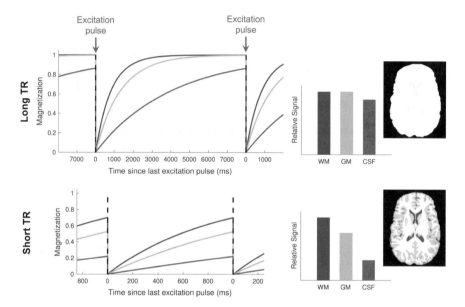

Fig. 11.2 Generating T_1 contrast by choosing the TR. Reproduced with permission from Short Introduction to MRI Physics for Neuroimaging, www.neuroimagingprimers.org

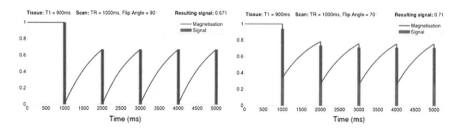

Fig. 11.3 Effect of flip angle on steady-state magnetization. Reproduced with the permission of Thomas Okell, Wellcome Centre for Integrative Neuroimaging

Exercise 11C Use the result from Exercise 10B to derive an expression for the Ernst angle: the optimal choice of flip angle for a tissue with a given T_1 value.

As we saw in Chap. 5, in reality, we get both T_1 recovery and T_2 decay—they are *simultaneous*, but *independent*, processes. Thus, we can choose what sort of contrast we want, as shown in Fig. 11.4, note that by 'short' or 'long' here we mean relative to the T_1 and T_2 values. We might imagine that we could combine their influence to help increase the contrast between different tissues by choosing an appropriate TR and TE. In practice, it is most common to get either a T_1-weighted or T_2-weighted image, something you will explore in Exercise 11D.

Notice that we only have the capacity to generate *weighted* images, i.e. these images are not directly the measurements of the T_1 or T_2 parameters themselves, but a tissue's T_1 or T_2 will affect whether it is bright or dark in the image. Notice also that we get proton-density-weighted images with a short TE and long TR; this is a combination where the images are relatively insensitive to both T_1 and T_2 and

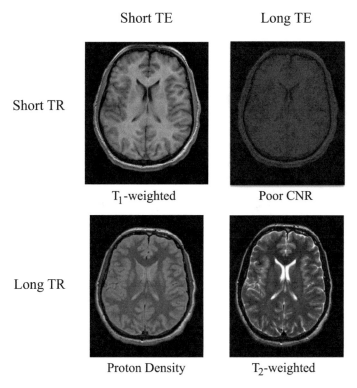

Fig. 11.4 Types of MR contrast and associated TR and TE required. Reproduced with permission from Short Introduction to MRI Physics for Neuroimaging, www.neuroimagingprimers.org

Table 11.1 Tissue properties for the tissue section from Exercise 2D

Tissue	T_1 (s)	T_2 (ms)
1	1.3	50
2	1.1	60
3	1.9	70

Table 11.2 Tissue properties for the tissue section in Exercise 3C

Tissue No.	Thickness (cm)	T_1 (s)	T_2 (ms)
gel	0.1	–	–
1	0.5	0.8	120
2	1.0	1.1	350
3	1.5	4.1	390

thus only depend upon M_0, the equilibrium magnetization, as we saw in Chap. 5 this depends upon the density of protons in the tissue.

Exercise 11D

(a) In Exercise 2D, the aim was to identify suspected cancerous tumours from tissue properties under X-ray imaging. If instead MRI was to be used, given the tissue properties in Table 11.1:

    (i) Would T_1- or T_2-weighted imaging produce the greatest contrast between tissue 2 and 3?

    (ii) Select a suitable repetition time (TR) and echo time (TE) that could be used.

(b) In Exercise 3C, the aim was to use ultrasound to image a section of tissue. If instead MRI was to be used, which pair of tissues would give the largest contrast with imaging parameters: TE $= 20$ ms, TR $= 1$ s, flip angle $= 90°$, if the tissues have the properties given in Table 11.2?

11.1.2 Diffusion

It is possible to make an MRI image sensitive to the diffusion of water by applying a gradient for a fixed length of time after excitation, whilst diffusion of water occurs, and then after a 180° pulse, the reverse gradient. Just as in the spin echo experiment, this process should result in dephasing of the signal (due to the gradient) and then rephasing, as long as the protons do not move. However, movement due to diffusion will cause different protons to experience a different field over time (as they move through the gradient). This will mean we will be unable to recover all of the dephasing that occurred and thus see an overall reduction in signal in direct proportion to the gradient strength and average distance moved (plus T_2^* effects).

The contrast generated by this process will depend upon the net diffusion of water, as well as the properties of the gradients. Typically, this is modelled as

$$S(T_E) = S_0 e^{-\frac{T_E}{T_2} - \gamma^2 G^2 D \delta^2 \left(\Delta - \frac{\delta}{3}\right)} \tag{11.2}$$

where D is the apparent diffusion coefficient (ADC). Notice that this includes the effects of T_2 decay which is then modified by a term that reflects the extra signal loss that occurs via the diffusion process. Greater diffusion, i.e. larger D, gives rise to more signal loss. The other terms are parameters associated with the diffusion acquisition. In body tissues, ADC will be lower than the diffusion coefficient for free water due to restrictions on diffusion imposed by the cellular content and cell walls. It is common to summarize the combination of different diffusion acquisition parameters in terms of the b-value

$$I(x, y) = I_0(x, y) e^{-b\, D(x,y)} \tag{11.3}$$

A diffusion-weighted image (DWI) thus contains contrast based on the ADC. A decrease in ADC has been observed in ischemic tissue that has died due to a lack of blood supply, and thus a DWI might be used to identify the extent of damage in someone who has suffered an ischaemic stroke. Whilst cell swelling occurs in ischemic tissue (called cytotoxic oedema), it is believed that the resulting reduction in the extracellular space gives rise to a reduction in ADC.

We noted in Chap. 10 that gradients can be applied in different directions in MRI, this being critical to producing a complete 2D or 3D image. The same is true for the gradients employed for diffusion contrast. For many tissues, this directionality is not very important. However, the brain is partly composed of the long axons of nerve cells that connect different parts of the brain to each other. These are generally grouped into bundles of 'fibres' in the white matter. The geometry of the axons means that water preferentially diffuses along the direction of the axon. By taking multiple diffusion measurements using gradients applied in different directions, it is possible to say something about the directionality of diffusion in a given voxel. The simplest example being the derivation of fractional anisotropy (FA), a measure of whether the diffusion is isotropic (FA $= 0$) or only in a single direction (FA $= 1$). More complete models of the full 3D pattern of diffusion, such as the diffusion tensor, can be calculated, which has led to the ability to 'track' connections between different parts of the brain by following the direction of diffusion along the white matter fibres.

11.2   Contrast Agents

If we cannot do something to the signal to achieve a greater contrast between the tissues we are interested in, then we might need to alter the properties of the tissues themselves to boost the contrast. This can, in various situations, be achieved through the use of a contrast agent. Typically, this is something that can be delivered by injection into the blood and is made of something that will alter the signal, such that it enhances the contrast of a particular tissue or structure in the body. Because contrast agents are often delivered in the blood, they are quite commonly used to enhance the contrast associated with the blood. For example, to image the vasculature, angiography; or to image blood delivery, perfusion. Most often, a contrast agent will be administered and then either imaged very rapidly or after sufficient time has elapsed for it to be fully delivered to its target. For example, for angiography, we need to capture the contrast agent whilst it is still circulating in the bloodstream. Whereas, for a targeted PET agent, we might need to wait long enough for the agent to have left the vasculature and accumulated in the tissue of interest. Very often, contrast agents are used in this way to produce a single image that captures the agent at a point of maximum contrast. But, as we will see in Chap. 12, it is also possible to use them dynamically to measure various physiological processes in the body.

11.2.1   X-ray Contrast Agents

If we want to create an X-ray contrast agent, we need something that will be very efficient at absorbing X-rays, which can be achieved using something that has a strong contribution from the photoelectric effect. Thus, like bone, we need to choose an atom that has a K-edge in the range of the X-rays we are using, typical examples are Barium (37.4 keV) or Iodine (33.2 keV). The former is used for imaging the gastro-intestinal tract and is administered orally as a 'Barium meal', whereas Iodine is delivered intravenously to image the vasculature.

11.2.2   Ultrasound Contrast Agents

If we want an ultrasound agent, we need something that creates a substantial reflection. Given that a gas/water interface produces a lot of reflected sound, due to the large difference in acoustic impedance, we exploit this effect to create an ultrasound contrast agent from bubbles. To make them compatible with the body, i.e. not likely to get trapped anywhere and cause a blockage in the blood supply, microbubbles of the order of 1 μm are used. Microbubble contrast agents are largely used for application in monitoring blood flow, such as in the heart. The reflections from microbubbles can get quite complex; it is generally simplest to regard them as scatterers, like the behaviour of red blood cells that we considered in Chap. 3. It is even possible for microbubbles to behave as little oscillators and produce harmonic signals.

11.2.3 SPECT/PET Contrast Agents

Both SPECT and PET are methods that inherently rely upon the introduction of a contrast agent—the radiotracer—since there is negligible inherent signal from body tissue. It is necessary to introduce the radionuclide into a molecule that can be safely introduced into the body and once introduced stays there for a sufficient length of time. This presents various challenges, but also the unique opportunity to 'label' molecules that are important in key chemical reactions that happen in the body, or that bind to particular sites within the body. There is thus a whole research field devoted to the creation of radiotracers with specific applications.

There are a wide range of radiotracers available for SPECT with an equally broad number of applications; many are based on Technetium-99 m (^{99m}Tc). A particular strength of SPECT imaging, apart from the relatively long half-life of the radionuclides compared to PET, is the ability to generate the tracer as needed on site, without needing access to a cyclotron. For example, ^{99m}Tc can be prepared from Molybdenum-99 (^{99}Mo), which is itself derived from a by-product of the fission of Uranium-235 in some power stations. The ^{99}Mo is delivered to a site in the form of a self-contained 'generator' and will produce a supply of ^{99m}Tc that can, via a sterile pathway, be incorporated into agents needed for imaging. Once the supply has reduced beyond what is usable, often after a week, the generator is swapped for another, and the original generator returned to a central facility to be replenished. In use ^{99}Mo converts to ^{99m}Tc within the column inside the generator, saline is drawn through the column to wash out the ^{99m}Tc. The ^{99m}Tc, which will be in the form of sodium pertechnetate, is then typically chemically modified using a specially prepared kit into a form suitable for a specific imaging application.

Whereas SPECT radionuclides tend to be transition metals and thus need careful formulation to be made useful for imaging applications, PET radionuclides include many of the more common species found in biological molecules or ones that can be substituted for common species such as fluorine for carbon. This means that it is possible to make PET-labelled molecules, the most common example being fluorodeoxyglucose (FDG), as shown in Fig. 11.5. This is very similar to glucose and the cells treat FDG as glucose, in that it is admitted across the cell membrane. However, it is unable to undergo the subsequent metabolic reactions and thus gets 'trapped' in the cells. FDG is a useful marker of cellular metabolism and the demands that the tissue has for glucose.

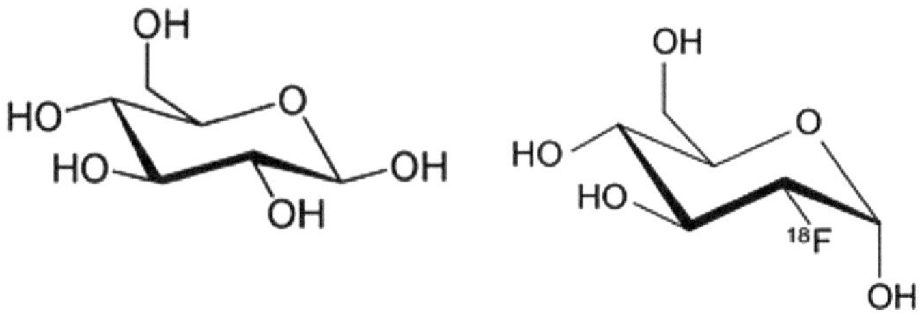

Fig. 11.5 Structure of glucose (left) and fluorodeoxyglucose (right)

11.2.4 MRI Contrast Agents

To create an MRI contrast agent, we need something that will manipulate the main MR contrast mechanisms, namely T_1 and T_2. Thus, we need a species that is inherently magnetic, the most common example being gadolinium (although it is possible to use iron too). Gadolinium is paramagnetic and its presence can alter both T_1 and T_2. When the gadolinium is in the blood, the magnetic gradient between the blood and tissue (without gadolinium) leads to more rapid dephasing and thus accelerated T_2 decay. If the gadolinium leaves the blood and accumulates in the tissue, this T_2 effect is reduced (as the gradient between tissue and blood is reduced) and instead a T_1 effect is seen. A common application is in the heart, where a gadolinium contrast agent delivered in the bloodstream is used to identify scar tissue after myocardial infarction (heart attack). In that application, the contrast agent accumulates preferentially in damaged tissue, this will show on a T_1-weighted image. The challenge with gadolinium is that it is toxic and thus has to be introduced into the body in the form of a chelate—essentially, the gadolinium atom has been to be enclosed in something to protect the body from it.

Further Reading
Further examples and applications of contrast manipulation and contrast agents in medical imaging can be found in

Introduction to Medical Imaging, Nadine Barrie Smith and Andrew Webb, Cambridge University Press, 2011.
Medical Imaging: Signals and Systems, 2nd Ed, Jerry L Prince and Johnathan M Links, Pearson, 2015.
Webb's Physics of Medical Imaging, 2nd Ed, M A Flower (Ed), CRC Press, 2012.

Further details on Magnetic Resonance Imaging contrast manipulation can also be found in
Introduction to Functional Magnetic Resonance Imaging, 2nd Ed, Richard B Buxton, Cambridge University Press, 2009.

Chapter 12
Beyond Contrast: Quantitative, Physiological and Functional Imaging

Abstract In this chapter, we will consider how we can use medical imaging to 'go beyond contrast' and simply the creation of 'pictures' for visual inspection, and instead how we can use medical imaging devices to make measurements. We will firstly explore how we can make medical imaging quantitative, so that image intensity values reflect a quantified measure of some physical property of the system. Then we will go on to explore how we can use medical imaging to measure functional and physiological, not just physical, properties of the body.

In Exercise 11A, you were asked to reflect on the physical processes that various imaging methods rely upon to generate contrast between tissues in the body. What you might have noticed is that almost invariably they are not quantities that are inherently physiologically interesting. Whilst it might be useful for the generation of a picture to exploit differences in acoustic impedance or T_1 relaxation time, it is often difficult to directly relate these quantities to more meaningful physiological quantities that would be expected to change in disease. Thus, if we want to get past imaging the 'structure' of tissues in the body, we need to exploit other processes or agents that respond directly to the physiology or function of the tissue.

We have already met contrast agents in Chap. 11 where our aim was to use them to generate or amplify the contrast in an image. This included the use of contrast agents that targeted particular physiological processes such as metabolism. While in Chap. 11 we used physiologically targeted agents to generate contrast, in this chapter we will consider how we can use dynamic measurements of the contrast agent to extract quantitative measures using tracer kinetics. We will also examine some endogenous mechanisms that we can exploit to gain insight into physiology. But first, we will consider how and when we can get quantitative images from medical imaging systems.

> **Exercise 12A**
> What challenges exist for the following imaging systems if we want to use them to make quantitative measurements of *physiology*:
> (a) Ultrasound

© Springer Nature Switzerland AG 2019
M. Chappell, *Principles of Medical Imaging for Engineers*,
https://doi.org/10.1007/978-3-030-30511-6_12

(b) PET
(c) MRI.

12.1 Quantitative Imaging

In Chap. 11, we considered how we might get more out of medical images than just the inherent information contained in the signals involved. But, in that chapter we were still only attempting to generate contrast in images: the intensities in the images themselves were not directly related to the quantity we were exploiting. For example, the intensities in a T_1-weighted image are not measurements of the T_1 relaxation time constant, nor are the values in an FDG-PET image measurement of the rate of uptake of FDG.

In theory, it should be possible to make quantitative measurements of the physical properties that we exploit to generate contrast in the images, such as attenuation coefficient or relaxation time constants. The advantage of making quantitative measurements is that we can potentially ensure that every image we produce will produce the same number consistently on whatever variant of the imaging device we use. And thus, we might be able to establish what values are 'normal' or 'abnormal' for a particular organ or when a particular disease is present. This can be valuable in identifying particular types of tissue that appear in disease, such as that associated with certain cancerous tumours.

For many cases, extracting quantitative measurements from images requires some form of calibration, so that we can interpret the intensity values in the images. An image, or set of images, from some standard object, often called a phantom, allows for the conversion of arbitrary intensity values into absolute numbers. However, the practicality of this varies from modality to modality.

12.1.1 Transmission Methods

In Chap. 2, we noted that for transmission methods, it is the attenuation coefficient that was the key physical quantity that gave rise to contrast. This coefficient is a tissue-specific property and thus can be used to not only distinguish between tissues, but actually identify the different tissues. In an X-ray transmission system, it is usual to quote the CT number, in Hounsfield units, and various tissues are associated with particular values. Calibration is required to control for variation in the source strength as well as detector sensitivity. Given that the CT number is a measure relative to the attenuation in water, a water-based reference object can be used to do the calibration.

For example, comparing the intensity in images generated with air only to that from a water filled object.

> **Exercise 12B**
> A CT scanner outputs a numerical value as a measure of the projection intensity at the detector in arbitrary units. Assuming a linear relationship between this value and the true intensity, derive a suitable calibration equation that converts the value into CT number when a value of 100 is measured with only air in the scanner, and 50 when a water phantom is scanned.

12.1.2 Reflection Methods

Reflection methods do not lend themselves to quantitative measurements. As we saw in Chap. 3, the signals in a reflection system depend upon the reflective index of the interfaces between different tissues. Even if we were to extract quantitative values, this would only relate to the interface and not immediately tell us specifics about the two tissues involved. There is the added complication of attenuation affecting the signal. We might attempt therefore to measure the attenuation coefficient to say something about tissue properties, akin to the CT number for X-rays. But this is complicated by scattering and distortion in the reconstructed image. The consequence is that quantitative measurements are not routinely attempted in ultrasound imaging, although some attempts have been made to interpret the speckle pattern in terms of properties of the tissue.

12.1.3 Emission Methods

In an emission system, the detector converts the received emissions, i.e. gamma photons in a SPECT or PET system, into a signal amplitude. Thus, if the properties of the detector are known it is possible to quantify the received intensity. However, to relate that to a quantitative measurement of emissions also requires corrections for attenuation and scatter, as well as any other processes that give rise to an increased background signal level, e.g. randoms in PET imaging. For PET, we saw in Chaps. 9 and 10 that CT images are often used to correct for attenuation: the CT-number values being converted into tissue type and a suitable typical attenuation value included in the reconstruction process.

The reason we might be interested in quantification for emission methods is so that we can measure the concentration of the tracer present in the voxel. For example, for FDG-PET we might wish to derive a quantitative measure of the metabolism in the tissue, which we would infer from the concentration of the tracer that had accumulated in the cells. This requires that the image intensity can be directly related to the concentration of the radiotracer and will involve calibration of the scanner using a source of known concentration.

Although calibration against a known source allows the tracer concentration to be measured, it is also necessary to account for other sources of variation that could affect the concentration observed in an individual patient. For this reason, it is typical to report the Standardized Uptake Value (SUV) to account for variability in the dose given and the patient size

$$SUV = \frac{I}{ID/BW} \qquad (12.1)$$

where I is the image intensity, ID the injected dose and BW the body weight. This equation normalizes the measured concentration by the concentration that would have been measured if the tracer has distributed evenly throughout the whole body, hence the ratio of the dose and the body weight. The calibrated image intensity will be in units of radioactivity per ml (radioactivity per unit volume); thus, the units of SUV are typically ml/g (in effect dimension-less).

In some cases, it might be more appropriate to measure the concentration relative to a reference region:

$$SUVR = \frac{SUV}{SUV_{ROI}} \qquad (12.2)$$

where SUV_{ROI} has been measured in a region of interest, normally defined somewhere on the image. In which case the dose need not be known, an internal reference is being used for calibration. Assuming this reference is chosen so that the measured intensity does not vary in disease, it allows for control of differences in normal radiotracer concentration between individuals meaning that deviation from normal can be detected.

12.1.4 Dynamic Data

In some cases, where the property of the system is a timing parameter, we might be able to derive quantitative values with needing calibration. A series of images taken over time (or under different conditions) might be acquired and the relative intensity in these images used to estimate the parameter. A good example of this are tracer methods, where we acquire so-called dynamic images, i.e. a series of images over time after introducing the contrast agent; something we will consider in more detail when we consider tracer kinetics in Sect. 12.2.

12.1.5 Resonance Methods

The other main example of exploiting timing information is MRI. As we have shown in Chaps. 5 and 11, in MRI the three main properties we can exploit to generate images are proton density, and T_1 and T_2 relaxation. In all cases, as we saw in Chap. 11, differences in these values can be used to generate weighted images through choice of TR and TE. For weighted images, the intensities are essentially arbitrary, related not only to a combination of these tissue properties but also properties of the MRI scanner itself. Thus, relating the image intensity back to any of these properties would appear to need some form of calibration, as has been necessary for other imaging modalities we have considered. However, both T_1 and T_2 are measures of the magnetization changing over time, and thus, it is possible to infer their values from a series of measurements, rather than trying to interpret the intensity on a single image. For T_2 quantification, we might use a number of images collected with different TE; for T_1 quantification, we could use a range of different TR or flip angles. There are some practical challenges to obtaining this information in a sufficiently short acquisition time; for example, when you change the TR, it takes a few repetitions for the signal to achieve steady state. It is also often important to account for variation in the B_0 and B_1 fields as these also have an influence on the image intensities. There is an increasing range of solutions that effectively sample at different TR, TE and/or flip angle and use this data to measure T_1 and T_2, as well as correct for field inhomogeneity.

12.2 Tracer Kinetics

A contrast agent can be used to increase the contrast of a given tissue or structure in an image, as we saw in Chap. 11. We might seek to relate the intensity of the agent in the image to aspects of physiology such as the rate of uptake of the agent from the blood, or available sites in the body to which the agent could bind. Beyond extracting information from an individual image, it is also possible to dynamically track the behaviour of a contrast agent over time as it passes through, or accumulates in, the imaging region. In these situations, we might call the contrast agent a tracer, and we can use the values from the image or dynamic series of images to extract quantitative (or semi-quantitative) information about physiologically related processes such as delivery or uptake/exchange using tracer kinetics.

Figure 12.1 shows a conceptual model of tracer kinetics, based on compartmental modelling principles. This is similar to the use of compartmental modelling in pharmacokinetics, with the main distinction being that tracer kinetics is concerned with the concentration of an agent in an imaging voxel rather than the whole body, or a whole organ. There are some limitations to compartmental models for tracer kinetics, not least that it ignores processes such as diffusion within the tissue and concentration gradients along blood vessels. However, it is generally sufficiently accurate for the length and time scales associated with many medical imaging applications.

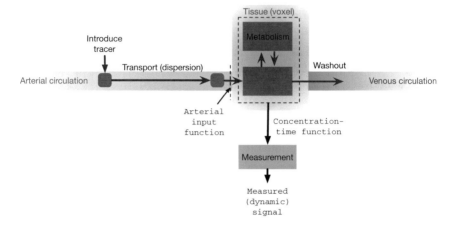

Fig. 12.1 A conceptual diagram of compartmental modelling for tracer kinetics

12.2.1 The Arterial Input Function

The tracer is present in some concentration in the blood (normally dissolved in the plasma) that varies with time, $C_p(t)$. The agent will be introduced somewhere upstream of the tissue of interest, for example, by intravenous injection, from where it will reach the arterial circulation eventually via the heart. The tracer will be transported in the circulation to the tissue. If we want to be able to calculate the delivery of tracer to the tissue, we need to know the dynamics of its delivery and how the concentration of the tracer varies with time; this is described by the arterial input function (AIF). For imaging, we are interested in the delivery of the tracer to a region of tissue—normally each voxel in turn—and thus we might use a different AIF for each voxel or assume that all voxels in the same tissue (approximately) share the same AIF.

If we are making dynamic measurements of the tracer and wish to extract quantitative measurements from them, it is important to know the AIF. One solution is to measure it by taking blood samples, which is the main solution for quantitative PET studies (but not used often in patients). However, that makes the procedure more time consuming and invasive. Given that we are already imaging the body, in some cases it is possible to extract the AIF from the images by looking at the time course in a voxel that is fully (or mostly) located within an artery. Alternatively, there are some cases where it is reasonable to assume the functional form for the AIF.

12.2.2 The Tissue (Voxel)

Once the tracer reaches the tissue, it might then leave the blood and enter the extravascular space, although some may remain only in the capillaries. The nature of the tissue, and what the agent is made from, will determine if remains outside of the cell, in the extracellular extravascular space (EES), or crosses the cell membrane and becomes intracellular.

Once it has arrived in the tissue, there are a number of routes via which the agent might be eliminated:

- **Washout**—the tracer leaves via the venous vasculature in the blood. If it entered the EES (and cells), it must diffuse back into the blood, normally due to the reverse of the concentration gradient as the agent is removed from the bloodstream elsewhere in the body, e.g. in the liver.
- **Metabolism**—the tracer may be metabolized inside the cell and converted to another substance, although this might prevent us from observing it with our imaging apparatus. Metabolism could involve a series of different reactions (both bi-directional and unidirectional) and thus may be modelled by a number of connected compartments.
- **Decay**—the agent may decay away via some physical process. We have already met such processes including the natural radioactive decay process for SPECT and PET agents.

The relevance of these different processes and compartments in the model depends upon the nature of the tracer itself. Importantly, our images most likely will not be able to distinguish between an agent when it is in the different compartments; thus, we only get a total measure of all the agent in the voxel, whether it is in the blood, the EES, the cells and also even if it has been metabolized in some cases. We may make some assumptions, such as the blood volume being far smaller than the tissue volume in most organs or that we do not need to distinguish between intra- and extracellular agent, to simplify our model.

12.2.3 A Simple Example: One-Compartment Model

The simplest example is of a one-compartment model. This is quite a reasonable approximation for a number of different contrast agents and assumes that the tracer is delivered to the tissue (extravascular, whether nor not it remains extracellular) by the blood and there is some process, physical (decay), metabolic or washout, that removes it at a fixed rate. We can write the model for this process as:

$$\frac{dC_t}{dt} = -k_e C_t(t) + F \cdot C_p(t) \tag{12.3}$$

where we have defined $C_t(t)$ at the concentration of the agent in the 'tissue' and we have an elimination constant k_e; $C_p(t)$ is the AIF, describing how the concentration of the agent varies with time on arrival to the tissue, we also have a constant, F, that defines the rate of delivery of the agent to the voxel of tissue.

At this point, we have to be quite careful about how we define concentration: C_t is the concentration of the agent in the tissue and thus will be per volume (or mass) of *tissue*. However, C_p is the concentration in the blood, so will be per volume (or mass) of the *blood* (or plasma). Thus, F will have units of volume tissue/volume blood/unit time. This quantity is perfusion: the delivery of blood to the tissue, typically measured in units of ml blood/100 g tissue/min. Thus, using a one-compartment model and having a measurement of the tissue concentration would allow us to measure perfusion in an organ.

Exercise 12C

For the one-compartment model in Eq. (12.3), use Laplace transforms (or otherwise) to find a time-domain solution for $C_t(t)$ for any arbitrary AIF $C_p(t)$.

12.2.4 The General Case: The Residue Function

Some tracers, particularly those that undergo a reaction once they enter the tissue, might require a multiple-compartment model to describe their behaviour, something we will consider further when we look at dynamic PET. There is a general form for the concentration of tracer that is applicable under any combination of compartments and tissue processes.

$$C_t(t) = F \cdot C_p(t) \otimes R(t) \tag{12.4}$$

where $R(t)$ is called the residue function. This is a generalization of the result from Exercise 12C: for the single-compartment model $R(t) = e^{-k_e t}$. The residue function describes what happens to a unit of tracer once it arrives in the voxel; thus, it has some particular properties determined by what is physically possible for the tracer to do:

- $R(t = 0) = 1$: this simply says that when a unit of tracer arrives it is all there;
- $R(t > 0) \leq 1$: this says that tracer either stays or is removed;
- $R(t)$ monotonically decreases[1]: this says that once tracer has been eliminated, it cannot come back again.

The residue function captures the behaviour of the tracer in the voxel. The form of the residue function might be an interesting thing to estimate in its own right

_____

[1]Formally, $R(t_2) \leq R(t_1)$, $t_2 > t_1$.

in situations where we do not already have a model for it. Where we want to explicitly separate the contribution from tracer in the vasculature from that in the tissue, a particular variant of Eq. (12.4) might be used:

$$C_t(t) = (1 - V_b)F \cdot C_p(t) \otimes R(t) + V_b C_p(t) \tag{12.5}$$

which includes a contribution from both tissue and blood weighted by the volume of blood in the voxel, V_b.

12.2.5 Deconvolution Analysis

In reality, the measurements of the concentration with time that we obtain from serial imaging data will be discrete. Like we did for the image formation/restoration and acquisition/reconstruction problems in Chaps. 6 and 10, respectively, we can write an equivalent matrix-vector version of Eq. (12.4)

$$C_t(t_j) \approx F \cdot \Delta t \sum_{i=1}^{j} C_p(t_i) R(t_j - t_i) \tag{12.6}$$

where we have approximated the convolution operation with a discrete summation assuming that the time series has been sampled evenly with spacing Δt, which is true for many (but not all) tracer kinetic applications. This gives

$$\mathbf{C_t} = F \cdot \Delta t \tilde{\mathbf{C}}_p \mathbf{R} \tag{12.7}$$

where $\mathbf{C_t}$ is the vector of measured concentration values, $\mathbf{R}$ a vector of residue function values and $\tilde{\mathbf{C}}_p$ is a matrix formed from the vector of AIF values $\mathbf{C}_p$ that implements the discrete convolution

$$\tilde{\mathbf{C}}_p = \begin{pmatrix} C_p(t_1) & 0 & \cdots & 0 \\ C_p(t_2) & C_p(t_1) & \cdots & 0 \\ \vdots & \vdots & \ddots & \vdots \\ C_p(t_N) & C_p(t_{N-1}) & \cdots & C_p(t_1) \end{pmatrix} \tag{12.8}$$

If we wish to solve for both F and $\mathbf{R}$ from some measured data, we can adopt a similar strategy as we did in Chap. 6 and perform deconvolution by inverting Eq. (12.7) using the matrix inverse of $\tilde{\mathbf{C}}_p$

$$F\mathbf{R} = \frac{1}{\Delta t} \tilde{\mathbf{C}}_p^{-1} \mathbf{C_t} \tag{12.9}$$

Given that by definition the residue function has $R(0) = 1$, we can estimate F as the magnitude of the first component in the vector $F\mathbf{R}$. Unfortunately, we meet the same issues as we had in Chap. 6 (and Exercise 6C) that this matrix is ill-conditioned, and matrix inversion thus amplifies the effect of noise in the measurements. Numerical schemes that avoid this often use a singular value decomposition of the matrix $\tilde{\mathbf{C}}_p$ as a means to filter out high-frequency (i.e. noisy) components.

12.2.6 Signal Model

The final aspect of tracer kinetic modelling is the measurement process itself. With imaging tracers, we are observing the *effect* of the tracer concentration within a voxel, not generally the concentration of the tracer itself. Depending upon the imaging method there might not be a simple linear relationship between the two and thus we need to include an observation or signal model. A particular limitation of most tracers is that we cannot distinguish them when they are in different compartments; we only measure the sum of all of the agents within the voxel. Thus, for a PET tracer, we cannot distinguish the different states of a metabolic tracer that undergoes reactions within the cell, which will limit how accurately we might be able to infer information about the rate of reaction for different reaction steps, i.e. the rate constant associated with the tracer moving from one compartment to another.

MRI with a Gadolinium agent is a good example of where we need to include a signal model within the tracer kinetic model. As we saw in Chap. 11, when using Gadolinium we do not observe it directly, but it changes the local T_1 and/or T_2 (T_2*) properties. We can convert from intensity in a T_1- or T_2-weighted image to concentration using the equations we met in Chap. 5 if we have a measure of the initial T_1 or T_2 values. Where these are not available, it is at least possible to convert to relative concentration and correct for the nonlinear relationship between signal intensity and T_1 or T_2 values.

12.3 Perfusion Imaging

Perfusion is a key measure of how many organs behave. Hypoperfusion (a reduction in perfusion) can have a very detrimental impact on organs such as the heart and brain. In the brain, hypoperfusion can be the result of either a blockage in a supply vessel or a bleed in a major blood vessel in the brain: these cause an ischaemic stroke or a haemorrhagic stroke, respectively. In the heart, hypoperfusion from the blood vessels supplying heart muscle can cause a heart attack. Since perfusion is so important to body organs, there are a range of imaging methods designed to measure perfusion.

As we saw in Sect. 12.2, perfusion plays a key role in the delivery of blood-borne tracers and thus appears as a key parameter in the simple one-compartment model accordingly. To measure perfusion, we typically need to examine the accumulation of a tracer in the voxel, either from a single image or from a dynamic series of images. Whilst the behaviour of a perfusion tracer often follows a one-compartment model, there is an important exception: the brain. Unlike the rest of the organs of the body, the brain is unusually restrictive as to what is allowed to pass from the blood into the tissue; the brain possesses what is known as the blood-brain barrier (BBB), which typically allows through only small molecules; most tracers are too large to pass into the brain tissue. For such tracers, this does not necessarily prevent perfusion imaging in the brain, but it means that the image needs to be interpreted differently.

12.3.1 Static Perfusion Imaging

A typical perfusion tracer would behave according to the one-compartment model in Sect. 12.2. Thus, an image acquired sometime after introduction of the tracer would have intensities related to the delivery of the tracer, i.e. perfusion, to each voxel. This would represent a perfusion-weighed image. The dynamic signal intensity from the one-compartment model also depends upon the elimination constant. For an ideal perfusion tracer, the elimination rate would be very slow compared to delivery. For many perfusion tracers, the only route for elimination is removal from the tissue via the bloodstream and thus the elimination constant is also related to the perfusion. Depending upon the imaging method, it may be possible to convert the image intensity into a quantitative metric via calibration.

In the brain, the BBB prevents most tracers from leaving the vasculature. Instead of image intensity being related to delivery of the tracer via exchange as in the one-compartment model, it is directly related to the volume of tracer and therefore blood, i.e. cerebral blood volume.

12.3.2 Semi-quantitative Dynamic Perfusion Imaging

An alternative to acquiring a single image of the tracer is to continually acquire as rapidly as the acquisition will allow from time of introduction of the tracer. It is then possible to examine the signal in each voxel arising from the tracer as a function of time. Depending upon how the tracer affects the image contrast, it may be possible to directly plot tracer concentration as a function of time, or after some conversion. From this time series, various semi-quantitative measures such as the Area Under the Curve (AUC) or Time to Peak (TTP) can be measured.

12.3.3 Perfusion Kinetic Imaging

Where it has been possible to obtain a dynamic imaging series of contrast concentration in a voxel, further quantitative metrics can be extracted if the arterial input function is also known. In some cases, it is reasonable to approximate the AIF using an analytical function. In other cases, the AIF might be measured from the images themselves, e.g. in the left ventricle of the heart for myocardial perfusion, or in a large cerebral artery for brain perfusion. For tracers that remain intravascular, it is possible to interpret the area under the concentration time series in terms of the volume of blood in the voxel by normalizing with the area under the AIF, normally called the Blood Volume (BV).

$$BV = \frac{\int_0^\infty C_t(\tau)d\tau}{\int_0^\infty C_p(\tau)d\tau} \tag{12.10}$$

Once the AIF is defined, an appropriate model can be fit to the data to estimate perfusion, most typically using a least-squares fitting method. An alternative is to use a deconvolution method. This approach is most often used for tracers that remain intravascular, since the residue function no longer represents the presence of the tracer in the tissue, but now describes the passage of the tracer through the vasculature. For this situation, it is possible to interpret the residue function in terms of properties of the vasculature in the voxel: the residue function describes how much of a unit tracer that arrives at one time still remains a given time later. For a voxel that contains a network of vessels, there are multiple different paths for the tracer through the voxel. Thus, the residue function represents the average effect of all of these different routes. The transit of tracer through the vasculature can be described in terms of a transit time distribution (TTD) $h(t)$, the corresponding form for the residue function is

$$R(t) = \int_t^\infty h(\tau)d\tau \tag{12.11}$$

This has the form of the (reverse) cumulative distribution for the tracer transiting the voxel. As we saw from Exercise 12D, performing deconvolution can be problematic and thus extracting the reduction function (let alone the TTD) with any accuracy is often not possible.

From the TTD, we might seek to characterize the vasculature in terms of its mean: the mean transit time (MTT). This can be a useful summary measure of the microvasculature, i.e. the capillary bed. Alternatively, with an estimate of the BV from comparison of the area under the concentration time curve and the AIF, and of the perfusion from deconvolution (often confusingly called the blood flow), it is possible to estimate the MTT without needing an accurate estimate of the residue function, via the Central Volume Theory

$$MTT = \frac{BV}{F} \tag{12.12}$$

12.3.4 Perfusion CT

As we saw in Chap. 11, vascular contrast agents for CT imaging are typically iod-
inated (i.e. contain iodine). A growing area of application of CT perfusion is in
stroke diagnosis, particularly determining the area of affected brain in a patient who
is experiencing an ischaemic stroke: arising from a blockage to one of the arteries
feeding the brain. Due to the BBB, the tracer does not leave the vasculature in a CT
perfusion scan; hence, a CT image of the tracer will reflect CBV, itself indicative
of tissue not being perfused, since if no blood is being delivered then no tracer can
arrive and thus there would be no signal.

12.3.5 SPECT Perfusion

It is typical to use a ^{99m}Tc-based tracer for SPECT perfusion, and this has applications
in brain perfusion, as well as in cardiac perfusion where two scans might be taken
firstly at rest and the whilst 'active', seeking to see how the perfusion changes when
the heart is put under stress, for example when someone is exercising. The 'active'
state is often induced by a pharmacological agent rather than making the patient to
perform strenuous activity.

12.3.6 Tracer Perfusion MRI

The added complexity for a Gadolinium tracer in perfusion MRI is that since it
affects T_1 and T_2, there is a nonlinear relationship between the measured change in
the image intensity and the concentration of the tracer, which has to be taken into
account in the compartmental model.

The main use of Gadolinium tracers in dynamic MRI, outside of the brain, are in
cancer where the highly vascular nature and leaky vessels in many tumours mean that
the tracer tends to accumulate preferentially in cancerous tissue (in the extracellular
space). The method relies on the T_1 changes induced by the tracer and is commonly
called dynamic contrast-enhanced (DCE) MRI. A simple two-compartment model
is typically used for DCE MRI.

As we have noted, the brain is slightly different to the rest of the body, because
the BBB prevents the Gadolinium tracer (amongst other things) leaving the blood
and entering the tissue. This means that the only effect observed is that of contrast
agent within the vasculature. In the dynamic susceptibility contrast (DSC) perfusion
MRI method dynamic T_2 (or T_2^*) imaging is used to capture the tracer as is passes
through the brain's vasculature. The result is that for DSC MRI, the passage of the
tracer through the vasculature results in a signal reduction as shown in Fig. 12.2. It

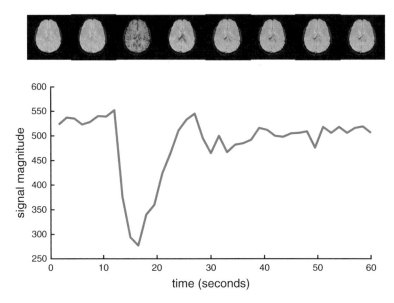

Fig. 12.2 Passage of a gadolinium tracer through the brain. The tracer remains in the vasculature, and this affects the tissue T_2, resulting in a reduction in the signal magnitude (as shown in a perfused voxel) as the tracer passes through the vessels in the voxel. Reproduced with permission from Introduction to Neuroimaging Analysis, Oxford University Press

is quite common to extract the cerebral blood volume (CBV) from DSC perfusion MRI and also MTT via deconvolution and the Central Volume Theory.

Typically, images can be acquired as often as once every one second, allowing deconvolution methods to be used. For kinetic quantification, an AIF is often measured in a large artery identified in the dynamic imaging data. Brain tumours exhibit the same leakage of tracer seen elsewhere in the body; thus, DSC MRI in brain tumours also has a contribution from extravascular tracer which complicates the interpretation.

12.3.7 Arterial Spin Labelling MRI

Arterial spin labelling (ASL) is a uniquely MRI method for measuring perfusion. Blood is 'labelled' in the neck using magnetic inversion of the water protons in the plasma. After a delay, to allow the labelled blood–water to flow into the brain and exchange into the tissue, an image is taken. This image will be a combination of accumulated labelled blood–water signal, plus a contribution from the (static) tissue. To remove the tissue contribution, a control image is acquired in the absence of labelling. The difference between this pair of images reveals a perfusion-weighted contrast. This process is illustrated in Fig. 12.3.

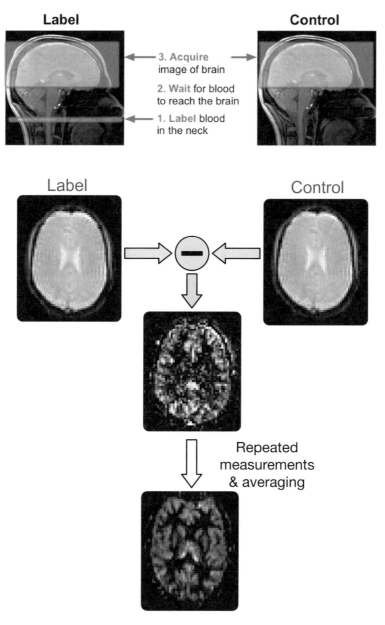

Fig. 12.3 Process of creating an arterial spin labelling MRI perfusion image. Reproduced with permission from Introduction to Perfusion Quantification using Arterial Spin Labelling, Oxford University Press

A simple one-compartment model is often used for ASL data, and the AIF is assumed to be known fairly accurately based on the labelling process. There is a direct linear correspondence between signal magnitude and the 'concentration' of the label, the constant of proportionality being the magnetization of the blood–water in the arteries in the neck, which can be estimated by a separate calibration scan. Thus, it is possible to extract quantitative measurements of perfusion using ASL. The main limitation of ASL is T_1 decay: once the blood has been labelled, the inversion of the protons returns to equilibrium with the T_1 time constant of the blood. This is in effect a route to elimination of the tracer as it reduces the apparent concentration, in this case though, the tracer signal is decreasing rather than a physical loss of tracer molecules from the voxel. The result of T_1 decay for ASL is that there are only a few seconds in which the blood can reach the brain and images be captured before the signal is lost.

Exercise 12D

A simple model for ASL MRI assumes that the labelling process creates an ideal AIF of the form

$$C_p(t) = \begin{cases} 0 & t \le \delta t \\ e^{-\tau/T_1} & \delta t < t \le \delta t + \tau \\ 0 & \delta t + \tau < t \end{cases}$$

where t is the time delay after labelling begins at which an image is taken, τ describes the duration of the labelling process, δt is the time it takes for the labelled blood–water to travel from the neck to a specific voxel in the brain, and T_1 describes the decay of the tracer that happens during its passage to the voxel.

Once the labelled blood–water has arrived in the voxel, it is then assumed that it remains there, the water having exchanged across the vessel walls from blood into the tissue. This gives a model for the residue function of

$$R(t) = e^{-t/T_1}$$

where the labelled water signal continues to decay with T_1 once it has been delivered.

(a) Derive and sketch the function that describes the signal arising from the labelled blood–water in a voxel of an ASL MRI experiment.

(b) For a given combination of δt and τ, at what time value would you sample to achieve maximum SNR?

(c) If, as is true in practice, δt is variable across the brain and between individuals, how might you choose a sampling time that would be insensitive to this parameter. What implications would this have for SNR?

12.4 Oxygenation

Since oxygen is key to the function of body cells, being able to measure its supply to tissues can be useful in disease, as well as monitoring changes in demand, such as those associated with neural activity in the brain. Perfusion is a critical process required in the delivery of oxygen and thus can be a useful proxy. However, normal perfusion does not necessarily imply normal oxygen delivery, or oxygen usage, if either the mechanism of oxygen transfer from the blood to cells, or cellular metabolism, is impaired. PET can be used to measure oxygenation, since it is possible to use ^{15}O as a radionuclide. However, as we saw in Chap. 4, it has a very short half-life, making it all but impractical for anything but research applications.

The blood oxygen-level-dependent (BOLD) effect describes how variations in blood oxygenation affect the local T_2^* of tissue. As we saw in Chap. 11, localized magnetic field inhomogeneities will affect how the MR signal decays, since different nuclei will have subtly different Larmor frequencies. This results in their phases going out of synchronization with each other and causes the T_2^* decay process. This process is enhanced by the greater variations in the field induced by deoxygenated haemoglobin (deoxyhaemoglobin), resulting in faster relaxation. In the case of increased metabolic demand in the brain, there is a reduction in the concentration of deoxygenated haemoglobin because the body (over) compensates for the demand by increasing blood flow and volume, as illustrated in Fig. 12.4. Thus, BOLD MRI uses T_2^*-weighted imaging to infer oxygen uptake in the brain, but not a spin-echo which would destroy the signal arising from the BOLD effect. BOLD has been most widely used to study brain activity by looking for regions of the brain where the

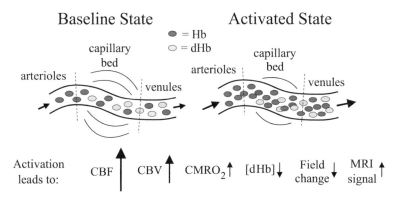

Fig. 12.4 Illustration of the BOLD effect: an increase in local neuronal activation (moving from the baseline state, left, to the activated state, right) is accompanied by large increases in perfusion (cerebral blood flow, CBF), cerebral blood volume (CBV), more modest increases in oxygen extraction (CMRO$_2$), and thus an overall increase in the amount of oxygenated blood (Hb) present compared to deoxygenated blood (dHb). As a consequence, the concentration of deoxyhemoglobin decreases, as do the magnetic field inhomogeneities that it causes, leading to an increase in the T_2^*-weighted MRI signal. Reproduced with permission from Short Introduction to MRI Physics for Neuroimaging, www.neuroimagingprimers.org

BOLD effect changes in response to the application of some form of stimulus such as watching something, making simple movements (e.g. finger tapping) or performing a task like memory recall.

12.5 Metabolic Imaging

12.5.1 Dynamic PET

In a PET study, we label something that is physiologically or metabolically interesting. Thus, the widespread use of FDG as a tracer because it reports on metabolic activity. Essentially, therefore, all the PET imaging we have met thus far can be considered metabolic. The use of semi-quantitative methods, such as calculating the Area Under the Curve, are applicable for dynamic PET generally, as they were for perfusion in Sect. 12.3.

Tracers such as FDG can often be modelled using the simple one-compartment model we met above. The fact that a PET tracer may undergo chemical reactions in the tissue (normally within the cell) means that we might need/want to build more complex compartmental models for some PET tracers that reflect the different products of those reactions. A generalized tissue model can be described by the following matrix (state space) equations:

$$\frac{d\boldsymbol{C}_{\mathbf{T}}(t)}{dt} = \boldsymbol{A}\boldsymbol{C}_{\mathbf{T}}(t) + \begin{bmatrix} K_1 & \mathbf{0} \end{bmatrix} \begin{bmatrix} C_{\mathrm{p}}(t) \\ C_{\mathrm{b}}(t) \end{bmatrix}$$

$$C_{\mathrm{t}}(t) = (1 - V_{\mathrm{b}})\mathbf{1}^{\mathrm{T}}\boldsymbol{C}_{\mathbf{T}}(t) + \begin{bmatrix} 0 & V_{\mathrm{b}} \end{bmatrix} \begin{bmatrix} C_{\mathrm{p}}(t) \\ C_{\mathrm{b}}(t) \end{bmatrix}$$

$$\boldsymbol{C}_{\mathbf{T}}(0) = \mathbf{0} \tag{12.13}$$

where $C_{\mathrm{p}}(t)$ is, as before, the AIF; in this case, it is specifically the concentration of the tracer in the blood plasma; $C_{\mathrm{b}}(t)$ is the whole blood concentration of the tracer and, with V_{b} the fractional blood volume accounting for the presence in the voxel of both tissue and blood compartments. K_1 is the influx constant for the tracer into the tissue from the blood plasma (into the 'first' tissue compartment). In this equation, $\boldsymbol{A}$ describes the rate of conversion between the different tissue compartments, which may be reversible processes. These compartments could represent steps in chemical reactions and movement of the PET tracer between different cellular components, e.g. into the mitochondria. Importantly, $C_{\mathrm{t}}(t)$, the measured concentration, is the sum of all the components present, hence the vector $\mathbf{1}^{\mathrm{T}}$.

A useful result is that the general solution to the equation above is of the form:

$$C_{\mathrm{t}}(t) = (1 - V_{\mathrm{b}})H(t) \otimes C_{\mathrm{p}}(t) + V_{\mathrm{b}}C_{\mathrm{b}}(t) \tag{12.14}$$

With $H(t)$ being composed of a weighted sum of exponential decay terms. You will see a simple example of this in Exercise 12E. This formulation has led to the 'spectral' method for PET kinetics. In that approach, instead of specifying the full compartmental model for a tracer and fitting that to the data, the data is fitted using this equation with a large number of exponential terms in $H(t)$ with fixed decay constants and the constraint that the weighting values should be non-negative.

More complex models for PET kinetics can be built. These might include the metabolism of the tracer in the blood, as well as the tissue. Additionally, it can be useful to consider two different tissues within the model so that the PET concentration data from one, 'reference' tissue, can be used in place of a measurement of the AIF, where this information is not available. In principle PET can, via kinetic models, provide a quantitative measure of tracer behaviour. However, a number of calibration and correction steps are required, including some we have already considered such as attenuation correction, as well as accounting for sensitivity of the detectors.

Exercise 12E

The tracer compartmental system in Fig. 12.5 describes the relationship between the concentration of radioactivity in arterial plasma, C_p, and that recorded by the PET camera in the tissue of interest, $C_t = (C_1 + C_2)$, for a particular irreversibly bound radiotracer.

(a) Write down the differential equations that describe the concentration in the two compartments.

(b) Hence show that the measured radioactivity can be related to that in the plasma by an equation of the form:

$$\mathrm{PET}(t) = K_1\left(\phi_1 e^{-(k_2+k_3)t} + \phi_2\right) \otimes C_p(t)$$

where you should determine the constants in the equation in terms of the rate constants in the model above.

Fig. 12.5 Example of a tracer compartmental system, for use in Exercise 12E

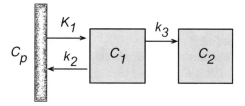

12.5.2 MRI

In Chap. 5, we deliberately concentrated on protons in water to generate a signal from within the body, noting that this was the most abundant source. In practice, there are a range of protons attached to other molecules; these have a slightly different Larmor frequency, and thus, it is possible to detect their contribution. However, because the abundance is much lower, the signal is proportionally smaller. Typically, such MR spectroscopic measures can only be made in relatively large volumes of tissue, or the order of centimetre-sized voxels, and thus any images are of a low spatial resolution, even if a range of different metabolites can be measured this way.

It is also possible to exploit exchange processes using MRI. For example, there is an exchange of protons between some molecules and water that we can use to create images that are specifically sensitive to metabolites or proteins that we could not normally detect using MRI or even using MR spectroscopy. This method also relies on the subtly different Larmor frequency of protons when in a different chemical environment (e.g. bonded to a carbon atom). For Chemical Exchange Saturation Transfer (CEST) MRI, protons are 'labelled' (normally using saturation) in that environment, and this label transferred to the water via exchange. If this exchange process is sufficiently rapid, we can accumulate enough 'label' on the water to measure a change even if the concentration of the other species is relatively low. In some cases, the exchange process is pH dependent, and this technique is being exploited to measure pH changes in the brain of stroke patients. The process is also dependent upon the relative concentration of the exchanging species, such as amide or amine groups on various molecules found in cells. This has been exploited in tumour imaging. Both of these applications have made use of exchange occurring naturally in cells without the need for any contrast agents, but it is also possible to develop specific agents that exhibit the same effect with the potential to make them more targeted, not unlike PET radiotracers.

Further Reading

Further examples and applications of quantitative and functional imaging can be found in

Webb's Physics of Medical Imaging, 2nd Edition, M A Flower (Ed), CRC Press, 2012.
Further details on functional and perfusion Magnetic Resonance Imaging can also be found in
Introduction to Functional Magnetic Resonance Imaging, 2nd Edition, Richard B Buxton, Cambridge University Press, 2009.
Introduction to Neuroimaging Analysis. Michael Chappell & Mark Jenkinson, Oxford University Press, 2018.
Quantitative MRI of the Brain, 2nd Edition, Mara Cercignani, Nicholas G Dowell & Paul S Tofts (Eds), CRC Press, 2018.

Appendix A
Exercise Solutions

A.1 Chapter 1

A.1.1 Exercise 1A

The point of this exercise was to have a look at medical practice, before having a lot of technical knowledge. Partly to appreciate that in practice there is more to the selection of imaging devices than just technical characteristics. But, also to get a sense for what sort of properties of imaging devices are important in clinical applications.

(a) Trauma Unit: Most probably some form of X-ray device would be used. This could be planar X-ray or even CT. One or both are often available in UK Accident and Emergency departments for example. Generally, their use will be motivated by the good contrast between bone and soft tissue, thus they are good for identifying bone fractures. The cost of X-ray imaging is generally lower than MRI and may call for less specialist knowledge, but MRI can be called on for soft-tissue damage that isn't easily seen with X-rays.

(b) Fetal: Ultrasound is widely used because it is non-invasive and non-ionising. It is a relatively low cost and increasingly highly portable imaging method. The information that can be gained from an ultrasound image can be limited particularly early in pregnancy when the foetus is small, and late in pregnancy when it is large. However, with a well-trained operator it is possible to measure critical information about the foetus allowing its health to be checked and its growth monitored. This is another application where MRI could provide more information, but is hugely challenging, you might find some examples of MRI research, but not clinical applications.

(c) Stroke: This is an example which varies geographically, often partially the method used is determined by what facilities can accommodate acutely ill patients. For example, in the UK CT is widely used as this is available in or near Accident and Emergency units, MRI units are generally not setup to care for such ill patients as those who are having a stroke. The purpose of imaging is firstly

© Springer Nature Switzerland AG 2019
M. Chappell, *Principles of Medical Imaging for Engineers*,
https://doi.org/10.1007/978-3-030-30511-6

to help distinguish between the two main type of stroke: ischemic and haemorrhagic. CT is okay for this as blood leaking into the brain (a sign of hemorrhage)
is fairly obvious. It is more challenging, and thus requires more expertise, to
identify from CT more information about the stroke, such as which areas of the
brain are affected in ischaemic stroke. MRI would be more favourable for this
as it is possible to examine changes in brain structure and physiology, it is also
more usually available for stroke patients in places like the USA.

(d) Doctor's Surgery: It is not common to find imaging devices in doctors surgery
or primary care locations. Non-imaging methods are often 'good enough' in the
first instance for a range of diseases and the use of imaging is a subsequent step
in more specialist diagnosis where it is needed. This also reflects the costs of
imaging devices as well as the specialist knowledge that is often called for to
use them, as well as interpret the images.

A.1.2 Exercise 1B

The fact that the expectation of a sum of random variables is a sum of the expectations
of the individual random variables implies that

$$E[\hat{y}] = \frac{1}{N} \sum_{i=1}^{i=N} E[y] = s$$

The fact that the variance of this sum is (for independent random variables) the
sum of the individual variances implies that

$$\text{Var}[\hat{y}] = \frac{1}{N} \sum_{i=1}^{i=N} \text{Var}[y] = \frac{\sigma^2}{N}$$

Thus, SNR on $\hat{y}$ will be

$$\text{SNR}_{\hat{y}} = \frac{s}{\sqrt{\sigma^2/N}} = \sqrt{N}\frac{s}{\sigma} = \sqrt{N}\,\text{SNR}_y$$

A.1.3 Exercise 1C

$$\text{CNR}_{AB} = \frac{C_{AB}}{\sigma} = \frac{|S_A - S_B|}{\sigma} = \left|\frac{S_A}{\sigma} - \frac{S_B}{\sigma}\right| = |\text{SNR}_A - \text{SNR}_B|$$

This shows that, as we might expect, that CNR depends upon the SNR and thus we wouldn't expect a method with poor SNR to deliver good CNR. However, a good SNR clearly doesn't guarantee a good CNR.

A.2 Chapter 2

A.2.1 Exercise 2A

(a) Intensity as seen by a detector located on the lines A and B:

$$I_A = I_0 e^{-\mu_1 \Delta}$$

$$I_B = I_0 e^{-\mu_1(\Delta - \delta) - \mu_2 \delta}$$

(b) Using the definition given (which we will meet again in Chap. 8):

$$\lambda_1 = \mu_1 \Delta$$

$$\lambda_2 = \mu_1(\Delta - \delta) + \mu_2 \delta$$

Hence

$$C = \lambda_1 - \lambda_2 = (\mu_1 - \mu_2)\delta$$

(c) This contrast depends on the relative attenuation coefficients and the thickness of the region of material with attenuation μ_2.

A.2.2 Exercise 2B

Total intensity, i.e., including scatter contribution, as seen by a detector located on the lines A and B:

$$I_A = N\varepsilon(E, 0)Ee^{-\mu_1 \Delta} + \left(\overline{\varepsilon_s E_s}\right)\overline{S}$$

$$I_B = N\varepsilon(E, 0)Ee^{-\mu_1(\Delta - \delta) - \mu_2 \delta} + \left(\overline{\varepsilon_s E_s}\right)\overline{S}$$

Note we have taken the scatter energy to be the same at both locations according to the approximation made in the text.

Relative contrast in measured intensity:

$$C = \frac{I_A - I_B}{I_A}$$

$$C = \frac{N\varepsilon(E,0)Ee^{-\mu_1\Delta}\left(1 - e^{-\mu_1\delta-\mu_2\delta}\right)}{I_A}$$

writing I_A in terms of the scatter-to-primary ratio:

$$I_A = N\varepsilon(E,0)Ee^{-\mu_1\Delta}(1+R)$$

$$C = \frac{1 - e^{(\mu_1-\mu_2)\delta}}{1+R}$$

TAKE CARE! It is not possible just to write I_A and I_B in terms of the scatter to primary ratio with exactly the same value of R and thereby cancel R.

If there were no scattering R would equal zero (and we would get back to the contrast we had in Exercise A). As the number of scattered photons increases so does R and this reduces C, and thus diminishes the contrast between the two detected signals. In turn, this would reduce our ability to distinguish between the two materials in the object, i.e., we would have reduced contrast between tissues in medical images.

A.2.3 Exercise 2C

Anti-scatter grids reject (absorb) scattered photons arriving at an angle incident to the primary photons. Hence their use will reduce R the scatter to primary ratio and thus increase the contrast.

A.2.4 Exercise 2D

The attenuation coefficient varies throughout the object, i.e., in both x and y:

$$\lambda(y) = -\log_e\left(\frac{I(y)}{I_0}\right) = \int \mu(x,y)\mathrm{d}x$$

This simplifies to a summation of the contributions from different regions as we consider all possible lines through the object between source and detector (i.e., as y varies).

Fig. A.1 A plot of the quantity λ as a function of y from exercise 2D

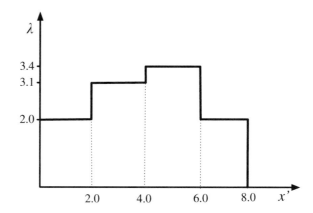

We need to calculate the attenuation coefficient from the given values of mass attention coefficient and density:

$$\mu_1 = 0.25\,\text{cm}^{-1}$$
$$\mu_2 = 0.6\,\text{cm}^{-1}$$
$$\mu_3 = 0.45\,\text{cm}^{-1}$$

Hence the sketch looks like (Fig. A.1).

Note that this is what in Chap. 8 we will call a single projection through the tissue. As we have noted in Chap. 2 this is effectively what we do in planar X-ray imaging. The relative contrast that can be measured this way between tissues 2 and 3 is

$$C = \left| \frac{\lambda_2 - \lambda_3}{\lambda_2} \right| = 0.97$$

The result is very similar if the other measurement is used as the denominator. Note that this implies that the contrast we can see this was is smaller than the relative difference of the attenuation coefficients, this is a result of the 'extra' attenuation due to the other tissue.

A.3   Chapter 3

A.3.1   Exercise 3A

An element of tissue, with constant cross-sectional area S, subject to a driving force, F, from the adjacent tissue.

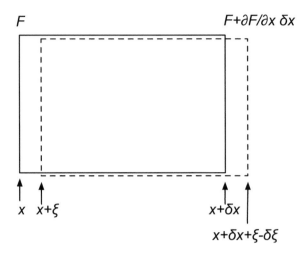

Newton's first law applied to forces on the element:

$$\left(F + \frac{\partial F}{\partial x}\delta x\right) - F = \rho_0 S \delta x \frac{\partial^2 \zeta}{\partial t^2}$$

Stress-strain relationship (using bulk elastic modulus, but noting that the cross-sectional area remains constant therefore volumetric strain is just 1D):

$$\frac{F}{S} = \kappa \frac{\partial \zeta}{\partial x}$$

Differentiate with respect to x:

$$\frac{\partial F}{\partial x} = \kappa S \frac{\partial^2 \zeta}{\partial x^2}$$

Combine to get:

$$\frac{\partial^2 \zeta}{\partial x^2} = \frac{1}{c^2} \frac{\partial^2 \zeta}{\partial t^2}$$

where $c = \sqrt{\kappa/\rho_0}$.

A.3.2 Exercise 3B

The general law for attenuation of intensity for ultrasound waves is:

$$I(x) = I_0 e^{-\mu x}$$

where μ is in units of nepers cm^{-1}. If we want the value in dB we need to convert to a base-10 logarithm, i.e.:

$$I(x) = I_0 e^{-\mu[\text{dB cm}^{-1}]x}$$

Selecting the transmitted intensity I_0 as our reference:

$$\mu[\text{dB cm}^{-1}] = -\frac{1}{x} 10 \log_{10} \frac{I}{I_0} = -\frac{1}{x} 10 \log_{10}(e^{-\mu x}) = 4.343\mu[\text{cm}^{-1}]$$

Intensity and pressure amplitude are related via:

$$I = \frac{p^2}{\rho c}$$

i.e., intensity is the square of pressure. If we take the square root of our expression for the general law of attenuation then:

$$p(x) \propto \sqrt{I(x)} = \left(I_0 e^{-\mu x}\right)^{\frac{1}{2}} = p_0 e^{-\frac{\mu x}{2}}$$

i.e., $\mu = 2\alpha$.

Following the same procedure as above:

$$\alpha[\text{dB cm}^{-1}] = -\frac{1}{x} 10 \log_{10} \frac{Q}{Q_0} = 8.686\alpha[\text{cm}^{-1}] = \mu[\text{dB cm}^{-1}]$$

i.e., the two attenuation coefficients take the same value when expressed as dB cm^{-1}.

A.3.3 Exercise 3C

We want intensity values relative to that transmitted, so set the incident intensity from the transducer as unity.

Firstly, calculate the reflection and transmission coefficients at each interface using the formulae in Chap. 3. Note that given the geometry, it is reasonable to assume that we can use the results given for waves arriving perpendicular to the boundaries.

$$R_{g1} = 0.0278, \ T_{g1} = T_{1g} = 0.9722$$
$$R_{12} = 0.004, \ T_{12} = T_{21} = 0.996$$
$$R_{23} = 0.0444$$

We need to account for attenuation. Using the information provided to calculate the attenuation coefficient:

$$\mu = 0.5 \times 2 = 1 \, \text{dB cm}^{-1} = 0.23 \, \text{cm}^{-1}$$

Thus for received intensity from the interfaces (being careful to account for attenuation on the transit through the tissues both in the forward and reflected parts of their passage):

$$I_1 = R_{g1}e^{-\mu \times 2 \times 0.1} = 0.0265$$
$$I_2 = T_{g1}R_{12}T_{1g}e^{-\mu(2 \times 0.1 + 2 \times 0.5)} = 0.0199$$
$$I_3 = T_{g1}T_{21}R_{32}T_{12}T_{1g}e^{-\mu(2 \times 0.1 + 2 \times 0.5 + 2 \times 1.0)} = 0.1993$$

Notice that the largest intensity in this case is associated with the deepest tissue interface despite intensity being lost through attenuation. The 'loss' of intensity due to reflections at the other interfaces is only a very small contribution.

We could also consider the reflections from interface between tissue 3 and the air on the far side of the object:

$$T_{23} = T_{32} = 0.9556$$
$$R_{3air} \approx 1$$
$$I_{air} \approx 0.0913$$

This will produce quite a strong reflection, but the intensity of this reflection is still smaller than that between tissues 2 and 3, despite the reflection coefficient being so high, due to attenuation.

The ultrasound intensity that reaches the far side of the tissue, i.e. is incident on the air interface is:

$$I_T = T_{g1}T_{21}T_{32}e^{-\mu(0.1 + 0.5 + 1.0)} = 0.2962$$

If we were using 8 MHz ultrasound this would reduce to 0.0054, a reduction by a factor of $e^4 = 54.6$ due to the linear dependence (in this case) of attenuation coefficient on frequency. This sample is only 3 cm thick, but attenuation has reduced the intensity of ultrasound by over two orders of magnitude during pass through the tissue (and it will be further attenuated as a reflection). Using ultrasound at this depth or even deeper would be challenging.

A.3.4 Exercise 3D

Starting with the definition of frequency in terms of speed c and wavelength λ:

$$f = \frac{c}{\lambda}$$

For the incident wave as 'seen' by the red blood cells who are moving the trans-ducer with speed $v\cos\theta$, the apparent speed of the wave $c^{\text{eff}} = c + v\cos\theta$, i.e., the arrival of the wave appears faster because the RBCs are moving toward it. Hence:

$$f_i^{\text{eff}} = \frac{c + v\cos\theta}{\lambda} = f_i \frac{c + v\cos\theta}{c}$$

Similarly, we can say for the wave received at the transducer that:

$$f_r^{\text{eff}} = f_r \frac{c + v\cos\theta}{c} = f_i \frac{(c + v\cos\theta)^2}{c^2}$$

Thus, the Doppler shift is:

$$f_D = f_i - f_r = f_i - f_i \left(\frac{c^2 + 2v\cos\theta + v^2\cos^2\theta}{c^2} \right) \approx \frac{2f_i v\cos\theta}{c}$$

where we ignore the term in v^2/c^2 because it is negligible when compared to v/c.

A.3.5 Exercise 3E

Using the formula gives 50 cm s^{-1}. Note that this is a fractional change of only 0.05%. If we wanted *velocity* rather than speed, we would need to know if this was a positive or negative shift in the frequency.

A.4 Chapter 4

A.4.1 Exercise 4A

This is just geometry:

$$R = \frac{d(L + 2z)}{L}$$

This equation is essentially a (simplified) measure of the resolution of the final imaging device and it tells us that objects that are deeper in the body can only be observed with poorer resolution than ones nearer the surface.

The role of the collimators in SPECT is to ensure that all detections are arriving from a well-defined direction—otherwise we have not hope of doing tomography i.e. reconstructing an image. Although the anti-scatter grid does something about scattered ray in the X-ray based method, for SPECT we are not reducing scatter since both scattered and genuine gamma rays might originate from anywhere in the body. In X-rays the source controls the direction the rays take and those that are scattered are normally 'knocked off course'.

A.4.2 Exercise 4B

Take a separation of 30 cm as being not a typical for different locations across the diameter of the torso. Using the same expression for attenuation as we have used in other chapters results in a difference in intensity of $\exp(-0.145 \times 30) = 0.0129$ over a distance of 30 cm^{-1}, i.e., a correction of up to 77.5. This would clearly not be insignificant when we are reliant on the intensity to tell us about the concentration of the radiotracer in different regions of the body.

A.5 Chapter 5

A.5.1 Exercise 5A

For a 1.5 T field strength the Larmor frequency is 63.87 MHz. At 3 T it is 127.74 MHz. Typically FM (Frequency Modulated) radio broadcasts are somewhere in the band 88–108 MHz (somewhat dependent upon the country in which they are based), but certainly close to the frequency of interest in clinical MRI systems.

A.5.2 Exercise 5B

$$\frac{d\mathbf{M}(t)}{dt} = \begin{pmatrix} \omega_0 M_0 \sin \alpha \ \sin(-\omega_0 t + \phi) \\ -\omega_0 M_0 \sin \alpha \ \cos(-\omega_0 t + \phi) \\ 0 \end{pmatrix}$$

$$\gamma \mathbf{M}(t) \times \mathbf{B}(t) = \begin{pmatrix} \gamma B_0 M_0 \sin \alpha \ \sin(-\omega_0 t + \phi) \\ -\gamma B_0 M_0 \sin \alpha \ \cos(-\omega_0 t + \phi) \\ 0 \end{pmatrix}$$

Hence $\omega_0 = \gamma B_0$ which is the Larmor frequency.

A.5.3 Exercise 5C

This is pretty trivial to show, but very worth understanding.

$$\mathbf{M}_{xy}(t) = M_0 \sin \alpha e^{-j(\omega_0 t - \phi)} = M_0 \sin \alpha \, \cos(-\omega_0 t + \phi) + jM_0 \sin \alpha \, \sin(-\omega_0 t + \phi)$$

M_x, the x-component of $\mathbf{M}$, matches the real part of this expression. Likewise M_y, the y-component of $\mathbf{M}$, the imaginary part.

A.5.4 Exercise 5D

Again, this is a fairly simple substitution, but worth understanding what is going on to get to this point. Following the transformation for x':

$$x' = x\cos\omega_0 t - y\sin\omega_0 t$$

Remembering that M_x is the x-component of $\mathbf{M}$ and M_y the y-component (and using angle formulae):

$$
\begin{aligned}
M'_x &= M_x\cos\omega_0 t - M_y\sin\omega_0 t \\
&= M_0 \sin \alpha \, \cos(-\omega_0 t + \phi)\cos\omega_0 t \\
&\quad - M_0 \sin \alpha \, \sin(-\omega_0 t + \phi)\sin\omega_0 t \\
&= \frac{M_0 \sin \alpha}{2}(\cos \phi + \cos(-2\omega_0 t) - \cos(-2\omega_0 t) + \cos\phi) \\
&= M_0 \sin \alpha\cos\phi
\end{aligned}
$$

We can do the same thing for M'_y and then apply the same complex notation we did in Exercise 5C.

A.5.5 Exercise 5E

Recall the equations of motion of the magnetisation vector

$$M_{xy}(t) = M_0\sin\alpha e^{-\frac{t}{T_2}}$$

$$M_z(t) = M_0\left(1 - e^{-\frac{t}{T_1}}\right) + M_0 \cos \alpha e^{-\frac{t}{T_1}}$$

Applying a flip angle of 90°, and assuming a phase angle of 0°, these equations reduce to:

Fig. A.2 A sketch of T_1 recovery and T_2 decay

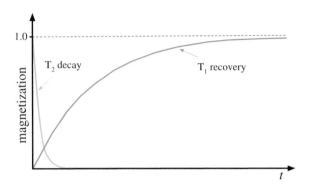

$$M_{xy}(t) = M_0 e^{-\frac{t}{T_2}}$$

$$M_z(t) = M_0\left(1 - e^{-\frac{t}{T_1}}\right)$$

A classic exponential decay curve for M_{xy} from a value of M_0 with time constant T_2, and a classic recovery curve for M_z toward M_0 with time constant T_1. A sketch should look something like (Fig. A.2).

A.6   Chapter 6

A.6.1   Exercise 6A

The voxel dimensions are $3.5 \times 3.5 \times 5$ mm in this case.

A.6.2   Exercise 6B

The PSF has a FHWM of 4.71 mm which is larger than the in-plane (x and y) size of the voxels from Exercise 6A, but smaller than the through plane (z) size (we might call this the 'slice thickness'). Thus, the in plane voxel size is a bit smaller than this 'true' resolution. But the z dimension is about right.

A.6.3 *Exercise 6C*

(a) Fairly trivially we just pre-multiply by the matrix-inverse of **H**

$$\hat{\mathbf{f}} = \mathbf{H}^{-1}\mathbf{g} = \mathbf{f} + \mathbf{H}^{-1}\mathbf{n}$$

This implies that we might attempt to recover the real object from the noisy imaging data, but it will still be corrupted by noise and we need to be concerned about the effect of the restoration process on the noise.

(b) We define

$$\text{SNR}_{\text{im}} = \frac{|\mathbf{f}|}{|\mathbf{n}|}$$

And likewise, using the result in part (a)

$$\text{SNR}_{\text{ob}} = \frac{|\mathbf{f}|}{\left|\mathbf{H}^{-1}\mathbf{n}\right|}$$

Using the result for the compatibility of matrix and vector norms

$$\left|\mathbf{H}^{-1}\mathbf{n}\right| \leq \left|\mathbf{H}^{-1}\right|\left|\mathbf{n}\right|$$

We arrive at the result given

$$\text{SNR}_{\text{ob}} \leq \frac{\text{SNR}_{\text{im}}}{\left|\mathbf{H}^{-1}\right|}$$

An ill-conditioned restoration operation will mean that the matrix inverse of **H** is ill-conditioned, resulting in a large value for the matrix norm, and hence a reduction in the SNR from the image to the restored object.

Note: we could have chosen any vector-norm for this result, but the 2-norm is often used practically to measure error in images.

(c) The quantity $|\mathbf{n}| = \mathbf{n}^{\text{T}}\mathbf{n}$ is the 2-norm of **n** and thus the squared error of the image compared to the true object

$$\mathbf{n}^{\text{T}}\mathbf{n} = (\mathbf{g} - \mathbf{H}\mathbf{f})^{\text{T}}(\mathbf{g} - \mathbf{H}\mathbf{f})$$

Differentiating and setting to zero gives

$$0 = -\mathbf{H}^{\text{T}}\mathbf{g} - \mathbf{g}^{\text{T}}\mathbf{H} + \mathbf{H}^{\text{T}}\mathbf{H}\hat{\mathbf{f}} + \hat{\mathbf{f}}\mathbf{H}^{\text{T}}\mathbf{H}$$

Which is satisfied by

$$\hat{\mathbf{f}} = \left(\mathbf{H}^{\mathsf{T}}\mathbf{H}\right)^{-1}\mathbf{H}^{\mathsf{T}}\mathbf{g} = \mathbf{H}^{-1}\mathbf{g}$$

(d) The derivation follows from part (c) with the matrix $\mathbf{Q}$ appearing in the final solution scaled by the multiplier τ. The role of $\mathbf{Q}$ is to 'condition' the matrix-inversion. By choosing it to be the identity matrix we are increasing the value of the diagonal elements of $\mathbf{H}^{\mathsf{T}}\mathbf{H}$, the matrix that is inverted as part of the reconstruction (see solution to part (c)). A purely diagonal matrix is always well-conditioned. Since in the image reconstruction case the matrix $\mathbf{H}^{\mathsf{T}}\mathbf{H}$ is poorly conditioned, this process will improve the condition and reduce the noise amplification. It will, however, result in a less faithful reconstruction, i.e., we will not be able to reduce all of the effects of the PSF (and may even introduce new artefacts). Various variants exist based on this idea, most of which will be more effective in practice than this formulation.

We have constrained the restoration by adding the term $\mathbf{Qf}$. The multiplier τ determines the weighting between minimising the error between the measured and restored image, and some property of the object defined by $\mathbf{Qf}$, were $\mathbf{Q}$ is simply an operation we have performed on the object (c.f. $\mathbf{H}$ that represents the image formation operation). We might choose $\mathbf{Q}$ such that it captures a particular property of the restored image, such as encouraging a spatially smoothed solution.

This is in the form of Lagrangian constrained-minimization problem, with τ a Lagrange multiplier.

A.6.4 Exercise 6D

These are standard functions and we can find their respective Fourier Transforms in standard tables. We are asked to consider what would happen if the imaging system had limited bandwidth, i.e., we didn't have high frequencies. Neither of these functions is inherently bandlimited, both have components at all frequencies. But, the Gaussian function (with a Gaussian-shaped transform) has diminishingly small components at high frequencies, so the loss would not be as bad as for the rectangular function (with a sinc-shaped transform). This comes back to the idea that sharp edges in the function, and thus in our images, requires high frequencies, as in the rectangular function. Whereas something smooth, like a Gaussian shape, doesn't. It will generally be more challenging to capture sharp edges in the image and in general these will get smoothed out, contributing to the Point Spread Function of the device and the effective resolution of the images.

A.7   Chapter 7

A.7.1   Exercise 7A

(a) Speed = distance/time. Hence a 1.3 μs echo has travelled 2 cm and a 6.5 μs echo 10 cm. Since this is the total distance travelled 'there and back', these interfaces must be at **1** and **5 cm** deep.

(b) Attenuation of an ultrasound wave is, for a uniform attention coefficient, given by:

$$I(x) = I_0 e^{-\mu x}$$

Thus, we need to correct by the factor TCG $= I_0/I$, the result follows using $x = ct$ to relate (total) distance travelled to echo time.

(i) $\mu = 3.439$ dB cm^{-1} = 0.792 cm^{-1}. For echo 1 (1.3 μs) TGC = **4.87**, echo 2 (6.5 μs) TCG = **2750**.

(ii) $\mu = 9.73$ dB cm^{-1} = 2.24 cm^{-1}. For echo 1 (1.3 μs) TGC = **88.2**, echo 2 (6.5 μs) TCG = **5.35 × 10⁹**.

This illustrates that using a higher frequency will limit depth penetration (even though it offers better spatial resolution).

A.7.2   Exercise 7B

If DoV is 10 cm then we need pulse to travel 20 cm, this takes 130 μs.
   Each frame takes 1/64 = 15.6 ms.
   Thus, number of lines per frame is 15.6/0.13 = **120**.

A.8   Chapter 8

A.8.1   Exercise 8A

As defined in Chap. 8:

$$\lambda_\phi(x') = \iint_{-\infty}^{\infty} \mu(x, y)\delta(x\cos\phi + y\sin\phi - x')dxdy$$

We have an attenuation profile of $\mu(x, y) = ax + b$.

When $\varphi = 0°$

$$\lambda_\phi(x') = (ax' + b)B$$

When $\varphi = 90°$

$$\lambda_\phi(x') = A\left(\frac{aA}{4} + b\right)$$

Note that the object is defined between x = $-A/2$ and A/2, y = $-B/2$ and B/2.

A.8.2 Exercise 8B

For SPECT the projection is given by:

$$\lambda_\phi(x') = \int_{AB} f(x, y)\mathrm{d}y'$$

i.e. it depends upon the line integral of the intensity, f, along the projection axis. We have already (for CT) re-written this in terms of a 2D integral, so we can write for SPECT:

$$\lambda_\phi(x') = \int\!\!\!\int_{-\infty}^{\infty} f(x, y)\delta(x\cos\phi + y\sin\phi - x')\mathrm{d}x\mathrm{d}y$$

The important feature of this equation is the $\delta(x\cos\phi + y\sin\phi - x')$ term, this will determine what we see if we were to plot x' versus ϕ. In this exercise we have a point source at (a, b), i.e. $f(x, y) = \delta(a, b)$. Hence the locus on the x' plot will be:

$$x' = a\cos\phi + b\sin\phi = \sqrt{a^2 + b^2}\sin\left(\phi + \tan^{-1}\frac{a}{b}\right)$$

Since the representation of a single point is sinusoidal when we plot x' versus ϕ, the representation of a collection of objects tends to look like it is composed of sinusoidal components, hence 'sinogram'.

A.9 Chapter 9

A.9.1 Exercise 9A

(a) Relationships between slice thickness and (mean) slice position:

$$z_1 = \frac{v_1 - \gamma B_0}{\gamma G_z}$$

$$z_2 = \frac{v_2 - \gamma B_0}{\gamma G_z}$$

$$\bar{z} = \frac{z_1 + z_2}{2} = \frac{1}{\gamma G_z} \frac{v_1 + v_2}{2} - \frac{\gamma B_0}{\gamma G_z} = \frac{\bar{v} - v_0}{\gamma G_z}$$

$$\Delta z = z_2 - z_1 = \frac{v_2 - v_1}{\gamma G_z} = \frac{\Delta v}{\gamma G_z}$$

(b) Re-arrange to get $\Delta v = \gamma G_z \Delta z = 42.58 \times 10^6 \times 40 \times 10^{-3} \times 5 \times 10^{-3} = 8.5\,\text{kHz}$.

  Note that the main field strength (3 T) doesn't affect this calculation, it would affect the choice of centre frequency though.

A.9.2 Exercise 9B

The Fourier transform of a rectangular function is a sinc function. To get a perfect slice selection we need a pulse that only has frequencies in a specific range, i.e. the bandwidth, and equal magnitude at all frequencies within the bandwidth. Thus, the pulse needs to generate a rectangular function in the frequency domain, this would require a sinc pulse in the time-domain. However, a sinc function has many side lobes and thus cannot be generated in a finite duration pulse. In reality careful shaping of the RF pulse will be required to get as precise a slice selection as possible.

A.9.3 Exercise 9C

Pulse sequence diagram and k-space trajectory for the following *readout* gradient (Fig. A.3).

$$v(x, y) = \gamma(B_0 + Gx + Gy)$$

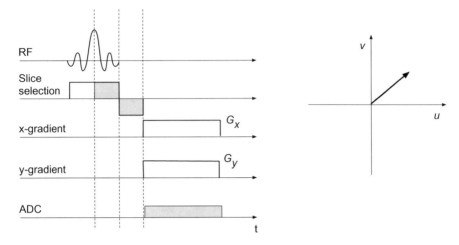

Fig. A.3 Pulse sequence diagram and k-space trajectory when equal magnitude *readout* gradients are applied in x and y simultaneously

A.9.4 Exercise 9D

Since readout takes some time, later values will have experienced more T_2 decay than earlier points. This will mean was we work through k-space later lines will be lower in magnitude according to an exponential decay process. This looks like a multiplication of the Fourier transform magnitude by an exponential decay, this is the same as a convolution of the true image with the FT of an exponential decay—short answer is that it will blur the resulting image.

The actual blurring effect will depend upon the duration of readout and the order in which we collect the k-space information. It is a particular problem for a full 3D readout where we try to get a full 3D k-space after a single excitation rather than working slice by slice.

A.10 Chapter 10

A.10.1 Exercise 10A

(a) $\text{COST}_{\text{SSD}} = \|\mathbf{\Lambda} - \mathbf{Pg}\| = (\mathbf{\Lambda} - \mathbf{Pg})^{\text{T}}(\mathbf{\Lambda} - \mathbf{Pg})$

(b) Differentiating the expression from (a) and setting to zero gives

$$\mathbf{\Lambda} - \mathbf{Pg} = 0$$

Being careful because $\mathbf{P}$ is non-square, we can find a solution to this using the pseudo-inverse as

$$g = (P^T P)^{-1} P^T \Lambda$$

(c) If **n** represents zero mean white noise and is thus a normal distribution with some (unknown) standard deviation (which is the same for all voxels) σ, then for N measurements

$$\text{Likelihood} = \prod_{i=1}^{N} P(\Lambda_i | g) = \prod_{i=1}^{N} \frac{1}{\sqrt{2\pi\sigma^2}} e^{-\frac{1}{2\sigma^2}(\Lambda_i - (Pg)_i)^2}$$

$$= \frac{1}{(2\pi)^{\frac{N}{2}} \sigma^N} e^{-\frac{1}{2\sigma^2}(\Lambda - Pg)^T(\Lambda - Pg)}$$

Note that this is written in terms of the image estimate **g**, rather than the true image **f**.

(d) Taking the (natural) log of the expression from part (c), we see that it contains a term that matches the SSD in part (a). Thus the maximum (log-) likelihood will occur when we have the solution as in part (b).

A.10.2 Exercise 10B

(a) This question is asking for the expression for the Poisson distribution

$$P(\Lambda_i | \langle \Lambda_i \rangle) = \frac{\langle \Lambda_i \rangle^{\Lambda_i} e^{-\Lambda_i}}{\Lambda_i!}$$

(b) If the counts are independent, then the total probability is the product of the individual LOR count probabilities

$$P(\Lambda | \langle \Lambda \rangle) = \prod_{i=1}^{N_{\text{LOR}}} \frac{\langle \Lambda_i \rangle^{\Lambda_i} e^{-\langle \Lambda_i \rangle}}{\Lambda_i!}$$

(c) Taking the (natural) log

$$\ln(P(\Lambda | \langle \Lambda \rangle)) = \sum_{i=1}^{N_{\text{LOR}}} \Lambda_i \ln(\langle \Lambda_i \rangle) - \langle \Lambda_i \rangle - \ln(\Lambda_i!)$$

In this expression Λ_i are the measured detections on the LOR and $\langle \Lambda_i \rangle$ are the predicted (expected) values. This means that $\langle \Lambda_i \rangle$ depends upon the estimated image, and thus the term $\ln(\Lambda_i!)$ will not vary during an iterative process where we vary the estimated image to maximise the cost function. Hence, we can ignore this term when defining the cost function.

A.11 Chapter 11

A.11.1 Exercise 11A

Physical properties that generate of contrast between tissues in the body:

a. CT—attenuation (interaction of photons with tissue).
b. Ultrasound—impedance differences between tissues.
c. MRI—proton density (but as we see later in the chapter we can also generate other contrasts based on physical properties such as T_1/T_2 decay constants and diffusion of water).

The point of this question is really to reflect on what imaging devices actually **measure**, making the point that they don't measure anything physiologically interesting directly—it is all physical quantities. Thus, they are useful for making images in many medical applications, but we might want to do something to create a more physiologically relevant contrast.

A.11.2 Exercise 11B

The longitudinal magnetization when we are starting at an initial condition $M_{z0} \neq M_0$ is:

$$M_z = M_{z0}\cos\alpha + (M_0 - M_{z0}\cos\alpha)\left(1 - e^{-\frac{t}{T_1}}\right)$$

This is a simple substitution—the point is that the expression given is the more general solution allowing us to solve for the magnetization after we have already been applying pulses for some TRs.

If we repeat the flip angle every TR we never return to the equilibrium M_0—but achieve a steady state recovery level something like this example (Fig. A.4).

If we are interested in the steady state value of M_z at the end of the TR (just before another flip angle) then $M_z = M_{z0}$, hence:

$$M_z = M_{z0}\cos\alpha + (M_0 - M_{z0}\cos\alpha)\left(1 - e^{-\frac{TR}{T_1}}\right)$$

$$M_{z0} = M_0 \frac{1 - e^{-\frac{TR}{T_1}}}{1 - \cos\alpha e^{-\frac{TR}{T_1}}}$$

This expression depends upon the T_1 of the tissue and says that in steady state different tissues will have different measured magnetization based on their T_1 value, thus we should be able to tell them apart.

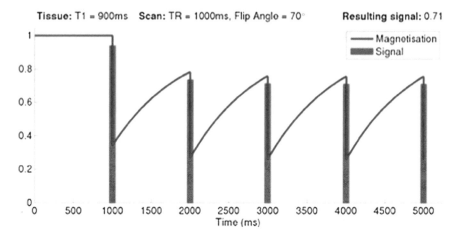

Fig. A.4 Example of reaching a steady state after multiple applications of a flip angle of 70°. Reproduced with the permission of Thomas Okell, Wellcome Centre for Integrative Neuroimaging

A.11.3 Exercise 11C

Remembering that we measure the transverse magnetisation M_{xy}:

$$M_{xy} = M_0 \frac{1 - e^{-\frac{TR}{T_1}}}{1 - \cos\alpha e^{-\frac{TR}{T_1}}} \sin\alpha$$

where we have used the result from part (c) as the magnitude of the component that we flip into the transverse plane. We want this value to be maximal for a given T_1 value by choosing α, so differentiate with respect to α and set to zero to get the Ernst angle: $\cos\alpha = e^{-\frac{TR}{T_1}}$.

A.11.4 Exercise 11D

(a) (i) Choose T_1-weighted imaging, since there are the largest (relative) differences in T_1 between tissues 2 and 3.

(ii) We need 'short' TR and TE relative to the T_1 and T_2 values respectively:

- TE $< T_2$—want TE to be as short as possible to maximise signal and thus SNR (and leave no effective T_2-weighting). A TE around 10 ms would be fine, we might struggle to get a lot shorter than this in practice (but 0 ms would be optimal).
- TR $< T_1$—need a short value compared to T_1, but this cannot be too short otherwise there will be no signal. Choose $T_1 \sim 1$ s.

(b) No calculation is required to answer this question:

- The TE is 'short' compared to the T_2 values.
- The TR is 'short' compared to the T_1 values.

Therefore, this will be a T_1-weighted image. Thus, we will be able to distinguish tissue 1 (or 2) from tissue 3. Tissues 1 and 3 will have the largest contrast because they have the biggest difference in T_1 (and T_2) values, this could be confirmed by calculation.

A.12   Chapter 12

A.12.1   Exercise 12A

Challenges for quantitative measurement:

d. Ultrasound—variability in speed of sound and attenuation appears on top of impedance differences and induces distortions. The whole system becomes non-linear and thus hard to calibrate.

e. PET—Attenuation variation, each photon will travel along a different path making correction tricky. We also need to control for instrumentation limitations, e.g. variability in detector sensitivity.

f. MRI—The images depend upon quantities such as T_1 and T_2 relaxation time constants which in themselves do not directly relate to physiology. We thus need to be able to relate physiological processes (or changes in physiological processes, e.g., in pathology) to changes in these fundamental parameters, or find other ways to manipulate the signal to make it sensitive to physiology; something that MRI is particularly good for.

The point of this exercise was to reflect on some of the challenges associated with being quantitative. A lot of imaging is about producing pictures, but if we want to make imaging more 'useful' with novel contrast agents or contrast mechanisms we might need to overcome some of these challenges.

A.12.2   Exercise 12B

If we assume a linear relationship as suggested, we expect something like

$$M = a\lambda + b$$

where M is the value we get out of the CT scanner for the projection intensity, λ is the projection intensity as defined in Chap. 8, and a and b are parameters we need to determine from the calibration data.

If we substitute using the equations in Chap. 2 for the relationship between intensity and attenuation coefficient and the definition of the CT number we get

$$M = a\mu_{H_2O}x\left(1 + \frac{CT}{1000}\right) + b$$

where we assume that the object (e.g., the phantom) has a thickness of x.

Using the data given we find that $b = 1000$ and $a\mu_{H_2O}x = -500$. This is sufficient to allow us to convert from the output of the scanner into a CT number in HU for the projection data, with that we can then reconstruct an image in HU.

A.12.3   Exercise 12C

Take the Laplace Transform of the equation:

$$sC_t(s) = -k_eC_t(s) + F \cdot C_p(s)$$

Note that we are assuming zero initial conditions, which is reasonable since we will generally be using a tracer that is not naturally occurring in the body. This does assume that we have been able to successfully extract only the time course for the tracer concentration from the image intensities.

Re-arrange:

$$C_t(s) = \frac{F}{s + k_e}C_p(s)$$

Note that this is in the form of a (first order) transfer function

$$C_t(s) = G(s)C_p(s)$$

With

$$G(s) = \frac{F}{s + k_e}$$

Using the property that multiplication in the s-domain is equivalent to convolution in the time domain, and recognising the standard form for an exponential function:

$$C_t(t) = F \cdot C_p(t) \otimes e^{-k_e t}$$

Which is of the form of an impulse response, with impulse response function $e^{-k_e t}$.

A.12.4 Exercise 12D

(a) We start with the equation for tracer kinetics in the general case

$$C_t(t) = F \cdot C_p(t) \otimes R(t)$$

and substitute in the expressions given. This gives rise to the expression for the tissue 'concentration' of

$$
C_t(t) = \begin{cases}
0 & t < \delta t \\
F T_{1b}\left(1 - e^{-\frac{t}{T_{1b}}}\right) & \delta t \leq t < \delta t + \tau \\
F T_{1b} e^{-\frac{t}{T_{1b}}}\left(e^{\frac{\tau}{T_{1b}}} - 1\right) & \tau \leq t
\end{cases}
$$

As an illustration of this solution consider the 'pcASL' solution in this figure (here $\delta t = 0$, $\tau = 1.8$ s, and $T_1 = 1.65$ s) (Fig. A.5).

(b) The maximum of this function occurs when $t = \delta t + \tau$ and this is the point where we would get maximum signal and thus expect the best SNR. In ASL,

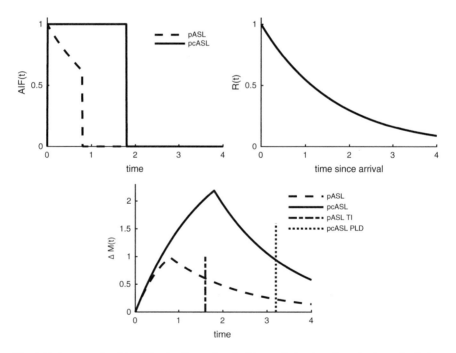

Fig. A.5 An example of ASL kinetics. Reproduced with permission from Introduction to Perfusion Quantification using Arterial Spin Labelling, Oxford University Press, 2018

we get to choose the value(s) of t at which we sample by choosing the delay between labelling and imaging.

(c) The point of maximum signal certainly depends on δt, and in fact a large part of the function will vary in magnitude with δt. The 'tail' is, however, insensitive to this parameter. Thus, as long as we sample after the peak of the function and choose a sample time that will satisfy this requirement for all possible values of δt we can get a measurement that depends on F without being affected by δt. The issue with this is that the signal magnitude is much lower, thus the SNR poorer. This is the motivation in the figure for choosing the sampling time indicated (pcASL PLD).

A.12.5 Exercise 12E

(a) Differential equations:

$$\frac{dC_1}{dt} = K_1 C_p(t) - (k_2 + k_3)C_1(t)$$

$$\frac{dC_2}{dt} = k_3 C_1(t)$$

$$\mathrm{PET}(t) = C_1(t) + C_2(t)$$

Note that the measured signal is of all of the radionuclide in the tissue, we cannot tell the difference based on which compartment it is in.

(b) Solving the differential equations using Laplace transforms yields:

$$sC_1(s) - C_1(0) = K_1 C_p(s) - (k_2 + k_3)C_1(s)$$

Since $C_1(0) = 0$

$$C_1(s) = \frac{K_1}{s + (k_2 + k_3)} C_p(s)$$

Using the convolution theorem

$$C_1(t) = K_1 e^{-(k_2 + k_3)t} \otimes C_p(t)$$

Also

$$sC_2(s) - C_2(0) = k_3 C_1(s)$$

Since $C_2(0) = 0$ and using the expression for $C_1(s)$

$$C_2(s) = \frac{K_1 k_3}{s(s + k_2 + k_3)} C_p(t)$$

Thus

$$C_2(t) = \frac{K_1 k_3}{k_2 + k_3}\left(1 - e^{-(k_2+k_3)t}\right) \otimes C_p(t)$$

Hence

$$\mathrm{PET}(t) = K_1\left(\phi_1 e^{-(k_2+k_3)t} + \phi_2\right) \otimes C_p(t)$$

With

$$\phi_1 = \frac{k_2}{k_2 + k_3},\ \phi_2 = \frac{k_3}{k_2 + k_3},\ \theta_1 = k_2 + k_3$$

Note that in this case K_1 is the rate of delivery, thus will be related to perfusion. The residue function for this system would be

$$R(t) = \phi_1 e^{-\theta_1 t} + \phi_2$$

Index

© Springer Nature Switzerland AG 2019
M. Chappell, *Principles of Medical Imaging for Engineers*,
https://doi.org/10.1007/978-3-030-30511-6

Printed in the United States
by Baker & Taylor Publisher Services